THE SUSTAINABLE MANAGEMENT OF VERTISOLS

THE SUSTAINABLE MANAGEMENT OF VERTISOLS

Edited by

J. KEITH SYERS

Naresuan University, Thailand

FRITS W.T. PENNING DE VRIES

International Water Management Institute (IWMI), Thailand*

and

PHIBION NYAMUDEZA

Chiredzi Research Station, Zimbabwe

with the assistance of
Robin N. Leslie (IWMI)*

CABI *Publishing*

In association with the
International Water Management Institute (IWMI)*

* In April 2001, the research program of the International Board for Soil Research and Management (IBSRAM) was merged into the International Water Management Institute (IWMI)

CABI *Publishing* is a division of CAB *International*

CABI Publishing
CAB International
Wallingford
Oxon OX10 8DE
UK

Tel: +44 (0)1491 832111
Fax: +44 (0)1491 833508
Email: cabi@cabi.org
Web site: www.cabi.org

CABI Publishing
10 E 40th Street
Suite 3203
New York, NY 10016
USA

Tel: +1 212 481 7018
Fax: +1 212 686 7993
Email: cabi-nao@cabi.org

A catalogue record for this book is available from the British Library, London, UK.

Library of Congress Cataloging-in-Publication Data
The sustainable management of vertisols / edited by J. Keith Syers, Frits Penning de Vries, and Phibion Nyamudeza ; with the assistance of Robin L. Leslie.
p. cm. -- (IBSRAM proceedings ; no. 20)
Includes bibliographical references.
ISBN 0-85199-450-4
1. Vertisols--Africa--Congresses. 2. Soil management--Africa--Congresses. 3. Vertisols--Congresses. 4. Soil management--Congresses. I. Syers, John K. (John Keith) II. Penning de Vries, F. W. T. III. Nyamudeza, Phibion. IV. International Board for Soil Research and Management. V. Series.

S599.5.A1 S86 2001
631.4--dc21 00-049358

Published in association with:

IWMI (International Water Management Institute)
PO Box 1025
10903 Jatujak
Bangkok
Thailand

This volume is published as IBSRAM Proceedings No. 20

ISBN 0 85199 450 4

Typeset in PhotinaMT by AMA DataSet Ltd, UK
Printed and bound in the UK by Biddles Ltd, Guildford and King's Lynn

Contents

Contributors

M. Adey, Department of Agriculture and Environmental Science, University of Newcastle upon Tyne, NE1 7RU, UK

Y. Ahenkorah, Department of Soil Science, University of Ghana, Legon, Ghana

E.K. Asiedu, Soil Research Institute, Academy Post Office, Kwadaso-Kumasi, Ghana

A. Astatke, International Livestock Research Institute, PO Box 5689, Addis Ababa, Ethiopia

A.T. Ayoub, PO Box 1488, Khartoum, Sudan

D.J. Beukes, ARC-Institute for Soil, Climate and Water, Private Bag X79, Pretoria 0001, South Africa

C. Bühmann, ARC-Institute for Soil, Climate and Water, Private Bag X79, Pretoria 0001, South Africa

A.D.C. Chilimba, Chitedze Agricultural Research Station, PO Box 158, Lilongwe, Malawi

R. Connolly, Queensland Department of Natural Resources, PO Box 318, Toowoomba, Queensland 4350, Australia

K. Coughlan, ACIAR (Australian Centre for International Agricultural Research), GPO Box 1571, Canberra, Australian Capital Territory 2601, Australia

E.T. Craswell, International Board for Soil Research and Management (IBSRAM), PO Box 9-109 Jatujak, Bangkok 10900, Thailand

N. Dalgleish, CSIRO, Tropical Agriculture, 120 Meiers Road, Indooroopilly, Brisbane, Queensland 4520, Australia

J. Deckers, Laboratory for Soil and Water, Catholic University of Leuven, Vitol Decosterstraat 102, Leuven 3000, Belgium.

M.C. de Villiers, ARC-Institute for Soil, Climate and Water, Private Bag X79, Pretoria 0001, South Africa

P. Drechsel, International Water Management Institute (IWMI) (formerly of International Board for Soil Research and Management, IBSRAM),

Regional Office for Africa, c/o University of Science and Technology, Kumasi, Ghana

D.M. Freebairn, Queensland Department of Natural Resources, PO Box 318, Toowoomba, Queensland 4350, Australia

T.J. Gerik, Texas A&M Experiment Station, Blackland Research Center, 808 E. Blackland Road, Temple, Texas 76502, USA

N. Hatibu, Department of Agricultural Engineering and Land Use Planning, Faculty of Agriculture, Sokoine University of Agriculture, PO Box 3003, Morogoro, Tanzania

J. Hussein, Department of Soil Science and Agricultural Engineering, University of Zimbabwe, PO Box MP 1267, Mt Pleasant, Harare, Zimbabwe

M.A. Jabbar, International Livestock Research Institute, PO Box 5689, Addis Ababa, Ethiopia

C.A. King, Department of Primary Industries, PO Box 102, Toowoomba, Queensland 4350, Australia

H.F. Mahoo, Department of Agricultural Engineering and Land Use Planning, Faculty of Agriculture, Sokoine University of Agriculture, PO Box 3003, Morogoro, Tanzania

Tekalign Mamo, Ethiopian Agricultural Research Organization, PO Box 2003, Addis Ababa, Ethiopia

A. Mapiki, Department of Agricultural Engineering and Land Use Planning, Faculty of Agriculture, Sokoine University of Agriculture, PO Box 3003, Morogoro, Tanzania

B. Matibiri, Zimbabe Sugar Association Experiment Station, Private Bag 7006, Chiredzi, Zimbabwe

N. Mukanda, Mt Makulu Central Research Station, P/B 7 Chilanga, Zambia

R.J.K. Myers, Natural Resource Management Program, International Crops Research Institute for the Semi-Arid Tropics, Patancheru 502 324, Andhra Pradesh, India

F. Nachtergaele, Soil Resources, Management and Conservation Service, Land and Water Development Division, Food and Agriculture Organization (FAO) of the United Nations, Viale delle Terme di Caracalla, 00100 Rome, Italy

P. Nyamudeza, Department of Research and Specialist Services, Chiredzi Research Station, PO Box 97, Chiredzi, Zimbabwe

J.W. Oteng, Agricultural Research Station, University of Ghana, Kpong, Ghana

P. Pathak, Natural Resource Management Program, International Crops Research Institute for the Semi-Arid Tropics, Patancheru 502 324, Andhra Pradesh, India

F.W.T. Penning de Vries, International Water Management Institute (IWMI) (formerly of International Board for Soil Research and Management, IBSRAM), PO Box 1025, Kasetsant University, Bangkok 10903, Thailand

K.N. Potter, USDA-ARS, Grassland Soil and Water Research Laboratory, 808 E. Blackland Road, Temple, Texas 76502, USA

M.E. Probert, CSIRO, Tropical Agriculture, 120 Meiers Road, Indooroopilly, Brisbane, Queensland 4520, Australia

F.B.R. Rwehumbiza, Department of Soil Science, Faculty of Agriculture, Sokoine University of Agriculture, PO Box 3008, Morogoro, Tanzania

M.A. Mohamed Saleem, International Livestock Research Institute, PO Box 5689, Addis Ababa, Ethiopia

H.P. Singh, Central Research Institute for Dryland Agriculture, Santoshnagar, Hyderabad 500059, India

O. Spaargaren, International Soil Reference and Information Centre, PO Box 353, 6700 AJ Wageningen, The Netherlands

G. Subba Reddy, Central Research Institute for Dryland Agriculture, Santoshnagar, Hyderabad 500059, India

J.K. Syers, Department of Natural Resources and Environmental Sciences, Naresuan University, Phitsanulok 65000, Thailand

A.J. van der Merwe, ARC-Institute for Soil, Climate and Water, Private Bag X79, Pretoria 0001, South Africa

M.C. Walters, ARC-Plant Protection Research Institute, Private Bag X134, Pretoria 0001, South Africa

Foreword

Black cracking clay soils, classified as Vertisols, are an important resource in the tropics and subtropics. In Africa, Vertisols occupy an area of over 100 million ha or 6% of the arable land area. These soils also feature significantly in the agricultural resource inventory of countries such as Australia, India and the USA. The physical properties of Vertisols make them difficult to cultivate and present inherent problems of low infiltration rates, waterlogging and high erodibility. In combination with widespread chemical fertility decline, the physical problems represent a major constraint to the sustainable management of Vertisol lands. Nevertheless, research shows that careful management of the soil surface can control and improve the soil water regime, significantly boosting crop yields. Work to overcome waterlogging with broadbed makers based on International Livestock Research Institute and International Crops Research Institute for the Semi-Arid Tropics work in Ethiopia has led to widespread adoption that has benefited poor farmers significantly. In Zimbabwe, at the drier end of the agroclimatic spectrum, the use of tied furrows has shown significant benefits to crop yields due to improved soil water conservation. On the Accra Plains in Ghana, the use of cambered beds has increased crop yields and fertilizer use efficiency. However, despite these success stories, the benefits of these new technologies are bypassing many millions of farmers cultivating Vertisols.

The International Board for Soil Research and Management (IBSRAM) took the initiative in organizing a workshop, held in Harare, Zimbabwe, in May 1999 to provide a forum for Vertisol researchers from all over the world to review their progress, learn from each other, and develop plans for collaborative research to fill any gaps identified. A key element in the choice of papers was the need for social and economic aspects of research to be considered alongside the biological and physical aspects. The workshop involved a field trip to Chisumbanje and a series of working groups designed to identify research needs. The output of the working groups is included in the synthesis chapter at the end of this book. Since the workshop, all chapters have undergone a process of peer review, editing, updating and revision.

The workshop and this publication would not have been possible without the support of the Zimbabwe Government. The Hon. K.M. Kangai, Minister of Lands and Agriculture, opened the meeting and officers of the Department of Research and Specialist Services (DR&SS) greatly assisted the field trip and workshop organization. Dr N.R. Gata, Director of DR&SS, Dr C.I. Muhambi, Head of Chemistry and Soil Research Institute, Dr P. Nyamudeza, and Mr J.K.R. Spurway deserve a special vote of thanks. We also appreciate the financial support for the meeting and for this publication provided by the Australian Centre for International Agricultural Research, Australian Agency for International Development, Danish International Development Agency and Department for International Development, UK. I also want to acknowledge the important back-up given to the meeting by our sister centre, the International Center for Maize and Wheat Improvement, through Dr Stephen Waddington and Mrs Rudo Shongedza.

The intention of this publication is to provide up-to-date information on Vertisols research and guide readers to important reference material. The forward-looking part on research needs indicates that much remains to be done before farmers, particularly those living in poverty, have the requisite tools and practices to cultivate Vertisols productively over the long term.

Eric T. Craswell
Director General
IBSRAM

Part I

Keynote and Overview Papers

Vertisols: Genesis, Properties and Soilscape Management for Sustainable Development

1

J. DECKERS,[1] O. SPAARGAREN[2] AND F. NACHTERGAELE[3]

[1]Laboratory for Soil and Water, Catholic University of Leuven, Leuven, Belgium; [2]International Soil Reference and Information Centre, Wageningen, The Netherlands; [3]Soil Resources, Management and Conservation Service, Land and Water Development Division, FAO, Rome, Italy

Key words: Africa, Vertisols, soil classification, soil physical properties, soil chemical properties, land use, farming systems, plant nutrition

Introduction

Vertisols are easily recognized because of their clayey texture, cracking structure and their dark colours. They can be highly productive but usually present management problems due to physical constraints.

Vertisols drew the attention of soil scientists as early as 1898 when the black soils covering a substantial part of peninsular India were recognized (Leather, 1898) because of their unique characteristics and importance for agricultural use. Later comparable black soils were also studied in other parts of the world, viz. in Eastern Gezira, Sudan (Dudal, 1965) and Texas, USA (Anderson and Byers, 1931).

Vertisols are notoriously variable and therefore require specific management. At the pedon scale the surface micro-variability of Vertisols imposes constraints on their use for agronomic research and agriculture in general. The inherent limitations of Vertisols are largely a function of the moisture status of the soils and the narrow range of moisture conditions within which mechanical operations can be conducted. The unique mineralogy of Vertisols makes these soils very susceptible to erosion, so that management practices must be geared to reduce soil loss (Eswaran and Cook, 1988). Temporal changes in physical attributes of these soils require accurate timing of agricultural practices.

This chapter presents a state of the art on our knowledge of Vertisols, their genesis and management in accordance with properties as defined in the World Reference Base (WRB) for Soil Resources classification (FAO *et al.*, 1998a) of soils with vertic properties (FAO *et al.*, 1998).

Local Names

Many local names exist for these soils. Dudal (1965, 1989) lists some 50 names, of which the best known ones are: *Regur* (India); *Adobe* (USA, Philippines); *Gilgai* (Australia); *Margalite* (Indonesia). Common names in Africa are: *Tirs* (Morocco); *Black Clays, Black Cracking Clays, Black Cotton Soils, Dark Clay Soils* (Anglophone East and South Africa); *Dian Pere* (Francophone West Africa); *Firki* (Nigeria); *Makande* (Malawi); *Mbuga* (Tanzania); *Mourcis* (Mali); *Badobes, Teen Suda* (Sudan).

Concept

Vertisols (from the Latin, *vertere*, to turn) are churning heavy clay soils which contain a high proportion of swelling clays such as smectites. When drying out, they form deep wide cracks from the surface downward at some period in most years, unless the soil is irrigated (Fig. 1.1).

The cracks usually show shear planes. At the surface of Vertisols a layer of fine granules ('grumic characteristic') may occur in varying thickness from several millimetres to a few centimetres. The upper part of the pedons commonly consist of strong and prism-like blocks. A *vertic horizon* occurs at some depth between 40 and 90 cm below the surface. It is a sub-surface horizon of a thickness of 25 cm or more, containing 30% clay or more throughout and which, as a result of shrinking and swelling, has either shear planes, or wedge-shaped structural aggregates with shiny and grooved curved surfaces ('slickensides'). The sliding of crumb surface soil into the cracks and the resultant shearing push the sub-surface soil upwards. In this way surface soil and sub-surface soil are mixed, a process known as 'churning' or (mechanical) 'pedoturbation'. In churning Vertisols, coarse fragments such as quartz gravel and hard, rounded, carbonatic nodules are concentrated at the surface, leaving the solum gravel-free. The coarse fragments are pushed upwards with the swelling soil, but most of the desiccation fissures that develop in the dry season are too narrow to let them fall back. As a consequence of the churning process, areas covered by soils with vertic properties often show a linear frequency of micro-knolls and depressions, collectively known as 'gilgai micro-relief' (Fig. 1.1).

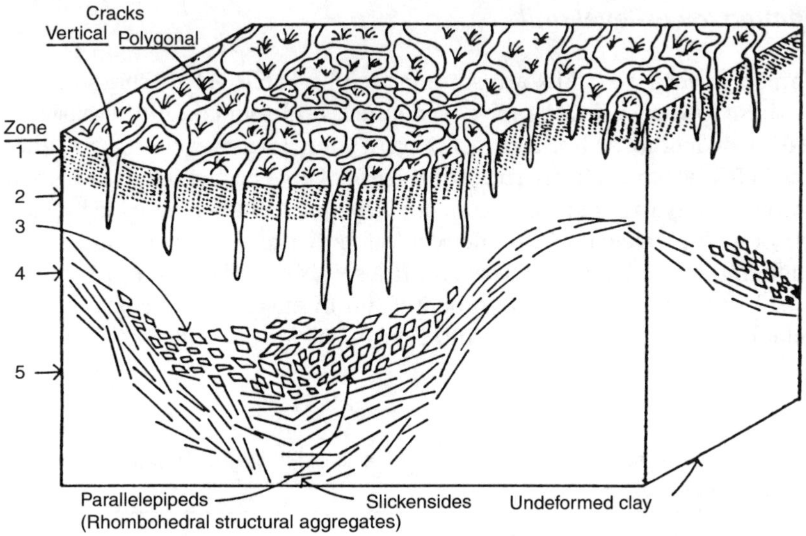

Fig. 1.1. Cross-section of a Vertisol pedon including gilgai micro-relief. (Source: Dudal and Eswaran, 1988.)

Soil morphological features seem to vary with varying climatic conditions (Jewitt *et al.*, 1979; Blokhuis, 1982; Dudal, 1989). With increasing rainfall:

- Cracks tend to be deeper and wider.
- The wedge-shaped soil structure is more distinct.
- Surface mulch weakens and becomes thinner.
- Crusting of the surface soil is more frequent.
- Organic matter in the upper horizons is higher and the colour becomes darker.
- The content of soluble salts decreases.
- Exchangeable sodium decreases or is absent.
- Reliability of rainfall improves.
- Hazards of flooding increase.
- Gilgais become more important.

Definition of Vertisols (FAO *et al.*, 1998)

Soils having:

1. A vertic horizon within 100 cm from the soil surface; *and*
2. After the upper 20 cm have been mixed, 30% or more clay in all horizons to a depth of 100 cm or more, or to a contrasting layer (lithic or paralithic contact, *petrocalcic*, *petroduric* or *petrogypsic* horizons, sedimentary discontinuity, etc. . . .) between 50 and 100 cm; *and*
3. Cracks that open and close periodically.

Common lower-level units

Common lower-level units are summarized in Table 1.1. It should be noted that these lower-level units may be defined, and named, on the basis of the presence of diagnostic horizons. At present three lower-level units are allowed in the WRB system. Priority rules for the use of lower-level soil names are to be followed strictly to avoid confusion. For example to classify a reddish-coloured Vertisol with a calcic horizon one would follow the priority list and note that qualifiers 6 and 12 apply. Therefore, the soil is classified as a Chromi-Calcic Vertisol. If more information on depth and intensity of the calcic horizon is available, e.g. occurring near to the surface, one may specify this by classifying

Table 1.1. Common lower-level units of Vertisols in World Reference Base for Soil Resources.

Qualifier	Summary description
Thionic	Having a sulphuric or sulphidic soil material within 100 cm from the soil surface
Salic	Having a salic horizon within 100 cm from the soil surface
Natric	Having a natric horizon within 100 cm from the soil surface
Gypsic	Having a gypsic horizon within 100 cm from the soil surface
Duric	Having a duric horizon within 100 cm from the soil surface
Calcic	Having a calcic horizon or concretions of secondary carbonates between 50 and 100 cm from the soil surface
Alic	Having a vertic horizon which has a cation exchange capacity equal to or more than 24 cmol(+) kg^{-1} clay throughout, a silt : clay ratio of less than 0.6, and an Al saturation of 50% or more
Gypsiric	Having gypsiric soil material at least between 20 and 50 cm from the soil surface
Pellic	Having in the upper 30 cm of the soil matrix a Munsell value, moist, of 3.5 or less and a chroma of 1.5 or less
Grumic	Having a surface layer with a thickness of 3 cm or more with a strong structure finer than very coarse granular
Mazic	Having a massive structure and hard to very hard consistency in the upper 20 cm of the soil
Chromic	Having a vertic horizon which in the major part has a Munsell hue of 7.5 YR and a chroma, moist, of more than 4, or a hue redder than 7.5 YR
Mesotrophic	Having a base saturation (by 1 M NH_4Oac) of less than 75% at 20 cm depth
Hyposodic	Having more than 6% saturation with exchangeable sodium in at least some sub-horizon more than 20 cm from the soil surface
Eutric	Having a base saturation (by 1 M NH_4Oac) of 50% or more at least between 20 and 100 cm from the soil
Haplic	Other Vertisols

the soil as a Chromi-Epicalcic Vertisol, indicating the occurrence of the calcic horizon within 50 cm from the surface (FAO *et al.*, 1998).

As these qualifiers refer to typical qualities which are important for the sustainable management of Vertisols, they will be discussed under the section 'Land use and management'.

Other soils with vertic properties

Vertic Leptosols, Vertic Solonchaks, Vertic Solonetz, Vertic Planosols, Vertic Chernozems, Vertic Kastanozems, Vertic Phaeozems, Vertic Gypsisols, Vertic Durisols, Vertic Calcisols, Vertic Luvisols, Vertic Cambisols. Some of the typical associations will be discussed under the spatial relationships.

Formation of the vertic horizon

Figure 1.2 places the most common clay minerals as separate fields at locations that qualitatively relate them to their formation requirements (Van Wambeke, 1991). The way the figure is drawn suggests that for given conditions only one mineral would form. In real soil solutions it is obvious that the chemical characteristics constantly change due to temporary variations in soil moisture, and that mixtures of clay minerals are synthesized. Their composition, however, reflects the most frequent set of soil solution attributes that are designated as clay formation environments (Van Wambeke, 1991). Figure 1.2 shows that smectites form in soil solutions rich in Si and Mg. The Si activity is close to or exceeds the activity in equilibrium with amorphous silica. Smectites, and consequently Vertisols, are either found in:

- semi-arid regions in depressions into which water seepage brings silica; or
- close to weathering basic rocks that are protected against leaching.

Favourable for the formation of smectite parent materials in which Vertisols form are (Driessen and Dudal, 1991):

- sufficient rainfall to enable weathering but not so high that leaching of bases occurs;
- dry periods for the crystallization of clay minerals that form upon rock or sediment weathering;
- impeded drainage that hinders leaching and curbs the loss of weathering products; and
- high temperatures that promote the weathering processes.

Under such conditions smectite clays can be formed in the presence of silica and basic cations – especially Ca and Mg – if the soil pH is above neutral.

The genesis of the vertic horizon centres around the formation of characteristic structural aggregates ('vertic structure') that may occur throughout

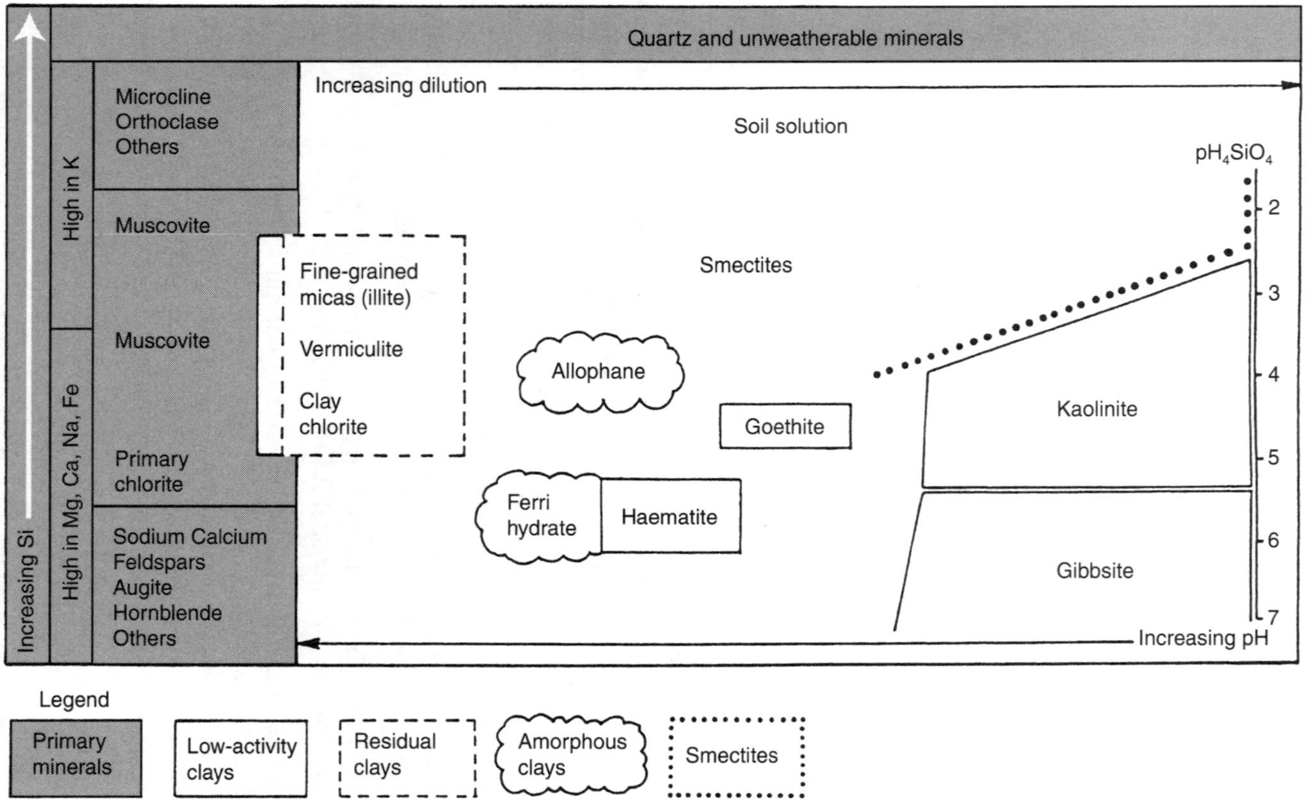

Fig. 1.2. Conceptual representation of clay formation environments. (Source: Van Wambeke, 1991.)

most of the solum but have their strongest expression in the vertic horizon; the aggregates change only gradually with depth in grade of development and size of peds (Driessen and Dudal, 1991).

When the saturated surface soil starts drying out, there is initially one-dimensional shrinkage, and the soil surface subsides without cracking. Upon further drying, the soil loses its plasticity; tension builds up in the soil material until its tensile strength is locally exceeded and the soil cracks. Cracks are formed in a pattern that becomes finer as desiccation proceeds. In most Vertisols, the surface soil turns into a 'surface mulch' with a granular or crumb structure. Vertisols that develop surface mulch are called 'self-mulching'. Granules or crumbs of the mulch fall into the cracks. Upon rewetting, part of the space that the soil requires for its increased volume is occupied by mulch material. Continued water uptake generates pressures that result in shearing: the sliding of soil masses along each other. Shearing occurs as soon as the 'shear stress' that acts upon a given volume of soil exceeds its 'shear strength'. Mass movement along oblique planes at an angle of 20–30° with the horizontal plane (Fig. 1.1) resolves the swelling pressure, acting in all directions. The shear planes are known as 'slickensides', polished surfaces that are grooved in the direction of shear. Intersecting shear planes define wedge-shaped angular blocky peds. The ped surfaces are slickensides or parts thereof, sometimes known as 'pressure faces'. The size of the peds increases with depth. The typical vertic horizon extends from some 15 or 20 cm below the surface mulch, down to the transition of solum to sub-stratum, just below the depth of cracking.

Distribution of Vertisols in Africa

Vertisols occur mainly in tropical and sub-tropical regions with a marked alternation of wet and dry conditions. Of the estimated 78 million ha of Vertisols in Africa, 69% occur in the semi-arid zones and the rest (21%) are found in the sub-humid zone (Deckers, 1993). Figure 1.3 (FAO, 1999) shows major occurrences in the Chad depression, in the Sud Depression of the Sudan, along the East African Rift Valley on basalt plateaus (Ethiopia, Kenya and Tanzania). Other African countries with occurrences of Vertisols are in order of importance: Zimbabwe, Somalia, South Africa, Nigeria, Ivory Coast, Ghana, Togo, Congo, Zambia, Burkina Faso and Madagascar. Vertisols are typically found in lower parts of the landscape such as river terraces, dry lake bottoms and other periodically wet areas.

Spatial and temporal relationships with some other reference soils

Vertisols are set apart from other soils by the combination of having a vertic horizon, a high clay content throughout, and deep, wide cracks upon drying. Other soils may show one or more of these properties, but never to the extent characteristic of Vertisols. They may have cracks not sufficiently wide, slickensides or wedge-shaped aggregates only, a vertic horizon underlying a coarser-textured surface layer, or they may be clayey but only have a vertic horizon that has not developed sufficient depth yet. These soils form intergrades and extragrades to Vertisols and are often associated with them in the landscape (Spaargaren, 1994).

The combination of topographic position, climatic conditions and parent material determines the spatial and temporal linkages of Vertisols with other soils. Vertisols normally occupy the lower parts of the landscape, comprising nearly level to gently undulating piedmont, flood and coastal plains.

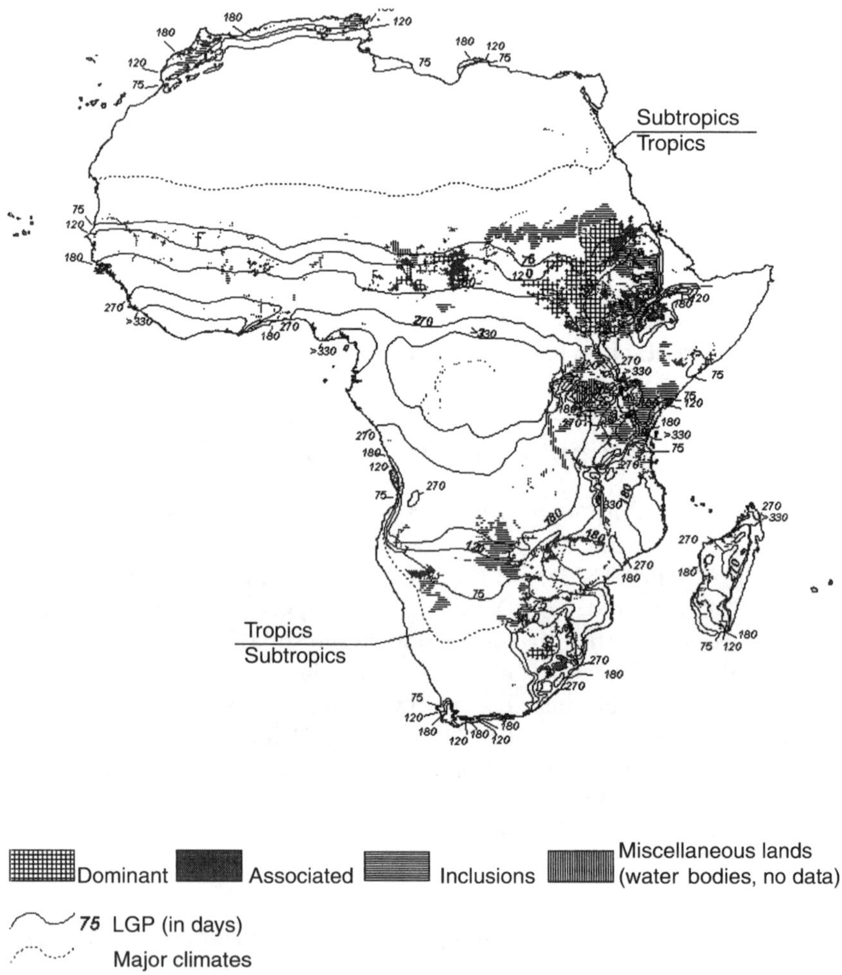

Fig. 1.3. Distribution of Vertisols in Africa and major climates and length of growing period (LGP) (based on WRB and the FAO/Unesco Soil Map of the World). (Source: FAO, 1999.)

Associated vertic intergrades (Vertic Calcisols, Luvisols, Cambisols) usually occur in relatively higher positions, comprising gently sloping to moderately steep plateau, mesa and pediment surfaces (Fig. 1.4).

In similar topographic positions Vertisols will merge on the arid side of the climatic spectrum into soils with accumulation of soluble components (Calcisols, Gypsisols, Solonchaks) due to the high evaporation. On the more humid side, accumulation of organic matter starts to prevail because of more abundant vegetation, giving rise to Phaeozems and Chernozems. In tropical and sub-tropical regions underlain by basic rocks toposequences with Nitisols/ Luvisols on the slopes and Vertisols/Planosols in low-lying positions occur frequently (Duchaufour, 1998). Sodium-rich parent materials are important in associations of Vertisols and Solonetz, with the latter soil in a transition position between prevailing upland soils (often Luvisols) and Vertisols. In riverine areas, depositional patterns play a role in the lateral linkages with other soils. Vertisols are often found in areas transitional to backswamps, associated with Solonetz, Planosols may occupy the more elevated positions, and Fluvisols, Gleysols (and even Histosols) central backswamps. In marine depositional environments, Vertisols may also be associated with Solonchaks.

Properties

Physically, Vertisols are very heavy clayey (30–95% clay) soils that become very hard and develop deep and wide cracks during the dry season. During the rainy season the cracks disappear while the land becomes fairly inaccessible due to a very slippery surface. They become sticky and plastic when wet and consequently traffickability is poor as moisture status is high. The apparent shrinking and swelling of the soil mass often result in small mounds and depressions at the surface, a phenomenon called 'gilgai'.

Although they have a relatively high water-holding capacity, shallow-rooting crops may suffer from drought stress. The most important physical characteristics of Vertisols are a low hydraulic conductivity, which seems to vary among the different Vertisols in line with salt content of the through flow

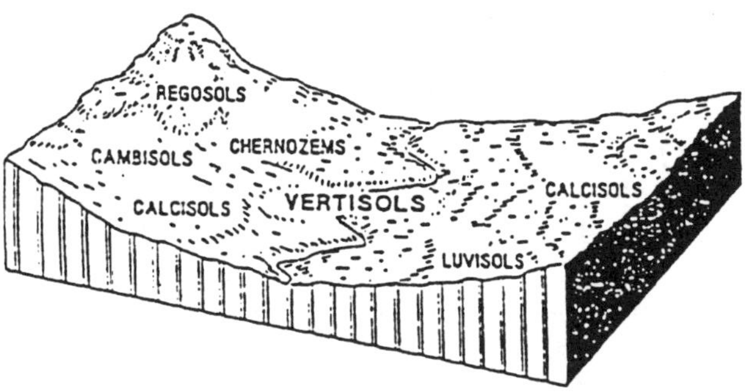

Fig. 1.4. Vertisol landscape with associated soils. (Source: Spaargaren, 1994.)

(Bouma and Loveday, 1988), a high infiltration rate when dry but a very reduced infiltration rate when wet, and a high bulk density (1.5–1.8 Mg m^{-3}). A major problem with Vertisols that urgently warrants research is gully erosion. Gully head formation is dominated by processes that are rather typical for soils with vertic properties such as rotational slumping on slickensides and sub-surface piping though the cracks.

Dudal (1989) stated that physical constraints to tillage and crop growth – plasticity, crusting, waterlogging and poor trafficability – are more pronounced in Vertisols of the humid zones.

Vertisols are relatively rich *chemically*, having a large reserve of weatherable minerals. Frequently they are dark coloured but have a moderate to low organic matter content. Vertisols generally have a high cation exchange capacity in the order of 30–80 cmol(+) kg^{-1} soil. The pH(H_2O) is neutral or slightly alkaline in most cases, but Vertisols with lower pH have been reported from the Mediterranean region. Base saturation is usually high, also because many Vertisols show accumulation of lime in some form or another. Dominant cations are Ca^{2+} and Mg^{2+}, while in places Na^+ plays an important role. In coastal regions Vertisols occur with high amounts of soluble salts and/or with sulphides or sulphates present. Salinity in Vertisols may also be caused by irrigation.

Mineralogically, Vertisols are characterized by the presence of large amounts of swell–shrink clay minerals. Most Vertisols have smectites, particularly montmorillonite, as the dominant mineral.

Soil *fauna* activity is very restricted in Vertisols. The only species that is reported frequently to occur is termites. Root activity is also reduced and root systems are normally confined to the cracks.

Land Use and Management

Though Vertisols form a great agricultural potential, they are difficult to work, being hard when dry and very sticky when wet. Many Vertisol areas in the semi-arid tropics still remain unused.

Farming systems on Vertisols

Agricultural use of Vertisols ranges from very extensive (rough grazing, firewood production, charcoal burning) through smallholder post-rainy season crop production (millet, sorghum, cotton, chickpeas) to small-scale (rice) and large-scale irrigated crop production (cotton, wheat, barley, sorghum, chickpeas, flax, noug (*Guizotia abyssinica*) and sugarcane). Cotton is known to perform well on Vertisols because cotton has a vertical rooting system that is not damaged too much by cracking. Tree crops are seldom recommended because tree roots find it difficult to establish themselves in the subsoil without being damaged by shrinking and swelling phenomena (Van Wambeke, 1991). Management practices for crop production ought to be directed primarily to controlling water dynamics besides maintaining or improving soil fertility.

Physical Vertisol land management

Physical constraints of Vertisols are related to their physical properties and moisture regime. Their heavy texture and the presence of expanding-type clay minerals result in a narrow range between moisture stress and water excess. Tillage is hampered by stickiness when wet and hardness when dry (Dudal, 1980). The agroclimatic potential of Vertisols is lower than other soil types because of their limited growing period. The single most important factor that reduces the available growing period below the estimates from climatic data is the waterlogging hazard in areas with surplus of rainfall over potential evapotranspiration (De Pauw, 1988). Because Vertisols have very low infiltration rates, excess water during the rainy season has to be drained and possibly stored in the soil for post-rainy season use ('water harvesting'). Several management practices have been devised to improve the water dynamics.

Evacuation of excess surface water

Surface drainage can be done by making broadbeds and furrows, as already practised by women farmers of Inewari village in the Central Highlands of Ethiopia (Jutzi *et al.*, 1987). This practice protects crops from waterlogging in the rooting zone. The drained water may be stored lower in the catchment in small ponds for other uses such as watering cattle, growing vegetables, etc. In order to relieve women of the painful drudgery of making beds and furrows by hand, ILCA developed the low-cost oxen-drawn broadbed and furrow maker (Jutzi *et al.*, 1987), which has been adopted by some 300,000 peasants of the Ethiopian Highlands. With reported yield increases of 150% with local wheat varieties and 300% on horse beans, there is no doubt that beds and furrows on Vertisols are successful. The only drawback of broadbeds and furrows recognized so far is increased soil erosion as a consequence of concentrated water flow in the furrows, causing rill and sometimes gully erosion. Broadbed and furrow technology solves problems on the individual farmer's field in isolation. Solutions still have to be found to bring the runoff water safely down to the lowest parts in the landscape (e.g. along grassed waterways), without causing severe soil erosion in neighbouring farmland. A participatory approach is needed, involving all the stakeholders, to solve this problem at the watershed scale.

Gully control

Control of gully erosion in Vertisols may require special dam constructions in the lower parts of the landscape, designed to keep up the groundwater table so as to maintain the subsoil moisture condition. In this way swell–shrink is inactivated and many processes related to gully formation (slumping, pipe erosion, subsoil cracking) are curbed.

Storage of excess water within the watershed

In the Tigray region, excess water from Vertisol soilscapes is harvested in micro-dams, allowing strategic irrigation of some 25 ha of Vertisols downstream of the dam site (Deckers, 1998a,b). Furthermore vegetables are grown throughout the year near the dam. Seepage losses from the dams usually benefit the ecosystems as a whole, since the water may surface as recharge in lower landscape positions. Livestock benefit from these micro-dams in many ways, e.g. by increased fodder availability from crop residues, presence of drinking water in the lake, fodder in low-lying recharge zones.

Though the micro-dam projects of Tigray generally are appreciated as a great success, two major problems are common: (i) salinization and sodification of the irrigation perimeters; and (ii) excessively high percolation losses. At some of the dam sites, up to 50% of the harvested water is lost per annum. This is a direct consequence of the swell–shrink of the smectite clay that may be present in the construction material. Possible solutions are use of a membrane or use of other construction materials, for example more weathered clay (e.g. from Nitisols that may occur in the same landscape).

Salinity buildup is a more serious problem in the sense that it has a direct impact on the lifetime of the micro-dams. Salinity may build up in a matter of a decade to such an extent that the whole dam has to be demolished and the surrounding land left to regenerate for several years before it can be cultivated again.

Water harvesting in Vertisol landscapes

Deep and wide cracking in dry Vertisols at the beginning of the growing period brings topsoil moisture levels in the available range at a later time than in other soils (De Pauw, 1988). Management should therefore be geared to storing as much water as possible in the subsoil, so as to build up available soil moisture reserves in order to extend the growing period beyond the estimated length of growing period from climatic data. Proven water harvesting techniques on Vertisols are:

- Small ponds may be used to harvest drainage water from individual farmer's fields, hence allowing water to be maintained high up in the watershed. This water can be used for strategic irrigation of vegetable gardens and for watering livestock (Jutzi *et al.*, 1988).
- Contour cultivation and bunding are used to improve infiltration. A beneficial side-effect of contour bunding is a check on soil erosion, which usually is a severe problem of Vertisols on slopes (Deckers *et al.*, 1998; Hurni, 1998). In the highlands of Northern Ethiopia continued contour ploughing has resulted in stepped landscapes (dagets), which may have a

height ranging from 0.3 m up to 3 m. Grasses occupy the riser and a more or less large strip on the shoulder (Nyssen *et al.*, 1999).

- Vertical mulching is sometimes practised to encourage infiltration in the subsoil. Crop stubble is therefore placed vertically in contour trenches with the stubble protruding 10 cm above the soil surface. Trenches are 4–5 m apart. Sorghum yields reportedly increased up to 50% by vertical mulching (Driessen and Dudal, 1991).
- In Zimbabwe farmers are putting the land under tied ridges in order to encourage water to infiltrate in the subsoil. It should be noted that this system will only be successful on strongly self-mulching Grumic Vertisols.

Improvement of rooting conditions

A number of special treatments are being practised to restore soil structure after many years of cultivation.

- In the Ethiopian Highlands a system of soil burning is done, called '*guie*', causing the clay fraction to fuse into sand-sized particles.
- Vertisols with indurated horizons within 100 cm depth (some cases of Calic and Gypsic Vertisols or Duric Vertisols) may benefit from a deep ploughing to break the hardened subsoil. The beneficial effect of deep cultivation and repeated ploughing on hydraulic conductivity and the oxygen diffusion rate was confirmed in Zambia (Mulonga *et al.*, 1997).

Plant nutrient management in Vertisols

Vertisols are considered the most fertile soils of the seasonally dry tropics (Duchaufour, 1998). Base saturation is high, with calcium and magnesium prevailing in the sorptive complex. Phosphorus availability is generally low. In many farming systems, land on Vertisols is left to fallow for 1–4 years to restore organic matter. However, increasing population has reduced this practice substantially and often land is only left fallow if it is completely degraded. Trials in many places have shown that continuous cropping may be sustainable with appropriate soil and water conservation and fertilizer management.

Extensive studies were made to correlate cotton yields with chemical and physical characteristics of the Vertisols in the Gezira scheme (Sudan), but results have remained controversial. Nachtergaele (1976) reported a statistically significant linear correlation with the organic carbon contents, which are generally low, and a parabolic relationship with the exchangeable sodium percentage.

Vertisols are usually deficient in nitrogen due to the generally low amount of organic matter. Nitrogen fertilizers have to be applied carefully in order to avoid losses through volatilization, denitrification and bypass flow (Sigunga, 1997). Not surprisingly N-response curves on Vertisols from many N-response trials show a near linear shape, since the efficiency of the applied nutrient is rather low (Deckers, 1988). Deep placement of nitrogen is best in dry years; in wet conditions, split banding is preferred (Van Wambeke, 1991). If supplied as ammonium, nitrogen may be capitalized on the exchange complex of Vertisols and therefore preserved against losses. Vertisols of the East African Highlands, developed on basalt, rarely show a strong response to phosphate under the low-intensive farming systems. When higher yields are obtained phosphate becomes an important limiting factor. The acid Alic Vertisols and some of the Chromic Vertisols contain high amounts of exchangeable aluminium and may be phosphate fixing. This can be remedied by application of lime. Other nutrients which may need correction are phosphorus and, occasionally sulphur and zinc.

The use of animal manure is in many cases limited by its competing use as fuel. So far trials are inconclusive on the possible advantages of animal manure in terms of overall soil organic matter or soil physical properties. Crop residues are used as animal feed, fuel and building materials rather than being returned to the soil (FAO, 1998).

Legumes have proven to increase crop yields of subsequent cereals remarkably as well as increasing the efficiency of mineral fertilizer uptake. Crop–livestock interactions can be enhanced by introducing a combination of broadbeds and furrows, phosphates, and undersowing cereals with legumes as a relay crop. The legumes will overgrow the cereal stover after harvest and hence become an improved live fodder for browsing, eventually enhancing animal draught power at the onset of the next ploughing season (Jutzi *et al.*, 1987; Gryseels, 1988).

Interpreting sustainable management of Vertisols from WRB lower soil units

Special care should be taken over salinity management of *Salic Vertisols*. In the case of the well-structured *Grumic Salic Vertisols*, deep ploughing followed by flushing according to the leaching requirement can easily leach excess salts. The *Mazic Salic Vertisols* are more impervious and need surface flushing at regular intervals. *Natric Vertisols* may occur in endoreic watersheds (e.g. in the Chad basin) and may be induced under poor irrigation management. Remedies comprise use of soil ameliorants such as organic matter and/or gypsum or any gypsum substitute. *Hyposodic Vertisols* build up sodium and require corrective action with respect to irrigation management. Sodification does not pose problems in the *Gypsiric Vertisols*.

Special cases are *Thionic Vertisols*, developed on old estuarine terraces. They contain very high amounts of acid sulphate in the subsoil, which have to be neutralized by liming or leached by base-rich irrigation water. This may prove rather difficult because of the slow infiltration rates of these soils.

Research Needs and Opportunities

In order to know why so many soils with vertic properties in Africa still produce below their potential, it is necessary to investigate research needs versus the opportunities for pay-off of investments in appropriate land management. Several situations occur:

- Many areas under Vertisols still remain underused because farmers face too many difficulties in solving the problems related to their physical management. Research in this case should focus on soil characterization, followed by quantitative land evaluation, in order to find the best match between sustainable land-use types and the agroecological and soil environment. In this analysis a holistic approach will be necessary that also will accommodate a non-use option, that is to preserve Vertisol areas as nature conservation areas.
- As discussed in the previous section, fundamental research is necessary to check the alarming rate of degradation of Vertisols presently under cultivation. Focus should go on better ways of gully control through 'inactivation' of swell–shrink, water harvesting systems versus surface evacuation of excess water. It is important that this research is conducted with a farming systems perspective, and on a hydrological watershed scale.
- Last but not least process research is needed on plant nutrient dynamics in Vertisols. Of particular importance are nitrogen efficiency and phosphate fixation in different types of Vertisols.

Conclusions

Although Vertisols make up a relatively homogeneous major soil group, they occur in a wide range of climatic conditions and show a considerable variability, which should be fully understood when developing technologies to improve their performance.

The climatic soil environment in Vertisols is drier than in other soils of the same area because, when dry, the evaporation surface is much larger due to deep cracks. When wet, Vertisols are the wettest soils in the soilscape and may show waterlogging due to low infiltration rates.

Most Vertisols are inherently fertile. Their physical management depends on the climatic zone, relief, and type of Vertisol. A sizeable number of low-cost technologies exist for improved physical and chemical Vertisol management, which have proved their value. However, most of these technologies were tested in isolation. A breakthrough may be expected if methods are devised to combine these technologies for the integrated watershed management of Vertisol-based farming systems. Crop–livestock interactions have proved to be positive in most of these systems. As most of the proposed Vertisol management technologies result in higher availability of fodder and drinking water throughout the dry season, livestock performance will be improved, which in turn will result in more timely field operations such as ploughing and sowing.

Despite their high potential, large tracts of Vertisols remain underexploited in sub-Saharan Africa. Any land clearing action however should be preceded by baseline mapping comprising soil inventory, land use, environmental significance and risks, potential farmers' problems and researchable issues. A participatory land evaluation involving all the stakeholders (including wildlife managers) will highlight the best strategy to target effective land management measures to come to optimal land resource use.

Acknowledgement

The authors would like to acknowledge the editors and Prof. R. Dudal for their valuable comments on this manuscript. Furthermore they wish to thank Ir. Kris Nackaerts (Laboratory for Forest, Nature and Landscape) for his valuable assistance in editing the Vertisol map of Africa.

References

Anderson, M.S. and Byers, H.G. (1931) *Technical Bulletin* No. 228. US Department of Agriculture, Washington, DC, USA.

Blokhuis, W.A. (1982) Morphology and genesis of Vertisols. In: Vertisols and rice soils of the tropics. Symposia papers 2. *Transactions of the 12th International Congress of Soil Science*, 8–16 Feb. 1982, New Delhi, India. Indian Agricultural Research Institute, New Delhi, India, pp. 23–47.

Bouma, J. and Loveday, J. (1988) Characterizing soil water regimes in swelling clay soils. In: Wilding, L.P. and Puentesm, R. (eds) *Vertisols: Their Distribution, Properties, Classification and Management*. SMSS Technical Monograph No. 18. Texas A&M University, College Station, Texas, USA, pp. 83–96.

Deckers, J. (1988) Soil fertility assessment of Ethiopian Vertisols on the basis of extension trial series of the Ministry of Agriculture. In: Jutzi, S., Haque, I., McIntire, J. and Stares, J. (eds) *Proceedings of a Conference held at ILCA*, International Livestock Centre for Africa, Addis Ababa, Ethiopia, 31 August to 4 September 1987, 431 pp.

Deckers, J. (1993) Soil fertility and environmental problems in different ecological zones of the developing countries in Sub-Saharan Africa. In: Van Reuler, H. and Prins, H. (eds) *The Role of Plant Nutrients for Sustainable Food Crop Production in Sub-Saharan Africa*. Dutch Association of Fertilizer Producers (VKP), Leidschenden, The Netherlands, pp. 37–52.

Deckers, J. (1998a) Ethiopia soil fertility management project. Background paper. *Consultants Report*, 38 pp.

Deckers, J. (1998b) 'Best bet land management options' to fit variability on farmers' fields in the Ethiopian highlands. *Proceedings Soil Fertility Management Workshop*, NFIA, Addis Ababa, Ethiopia, p. 30.

Deckers, J.A., Nacthergaele, F.O. and Spaargaren, O.C. (eds) (1998) *World Reference Base for Soil Resources.* Introduction. ISSS–ISRIC–FAO, Acco Leuven, 165 pp.

De Pauw, E. (1988) Assessing the agro-climatic potential of Vertisols. Management of Vertisols in sub-Saharan Africa, In: Jutzi, S., Haque, I., McIntire, J. and Stares, J. (eds) *Proceedings of a Conference held at ILCA,* International livestock Centre for Africa, Addis Ababa, Ethiopia, 31 August to 4 September 1987, 431 pp.

Driessen, P.M. and Dudal, R. (1991) *The Major Soils of the World.* Lecture notes on their geography, formation, properties and use. Agricultural University Wageningen, and Katholieke Universiteit Leuven, 310 pp.

Duchaufour, Ph. (1998) *Handbook of Pedology.* Soils – Vegetation – Environment. Masson, Paris, 264 pp.

Dudal, R. (1965) Dark clay soils of tropical and subtropical regions. *FAO Agricultural Development Paper* No. 83. FAO, Rome, Italy, 161 pp.

Dudal, R. (1980) Soil-related constraints to agricultural development in the tropics. In: *Soil-related Constraints to Food Production in the Tropics.* IRRI and Cornell University, Los Banos, The Philippines, pp. 23–37.

Dudal, R. (1989) Vertisols of subhumid and humid zones. *Proceedings IBSRAM Inaugural Workshop,* ICRISAT, India, pp. 55–60.

Dudal, R. and Eswaran, H. (1988) Distribution, properties and classification of Vertisols. In: Wilding, L.P. and Puentes, R. (eds) *Vertisols: Their Distribution, Properties, Classification and Management.* SMSS Technical Monograph No. 18, Texas A&M University, College Station, Texas, USA, pp. 1–22.

Eswaran, H. and Cook, T. (1988) Classification and management-related properties of Vertisols. In: Jutzi, S., Haque, I., McIntire, J. and Stares, J. (eds) *Proceedings of a Conference held at ILCA,* Addis Ababa, Ethiopia, 31 August to 4 September 1987, 431 pp.

FAO (1998). Ethiopia soil fertility initiative concept paper. *Report No. 98/028 CP-ETH.* FAO, Rome, 34 pp.

FAO (1999) Soil map of the world. Africa. *FAO Geo-referenced Database based on WRB and FAO/Unesco Soil Map of the World,* Vol. VI, Africa. FAO, Rome.

FAO–ISRIC–ISSS (1998) *World Reference Base for Soil Resources.* World Soil Resources Reports No. 84, FAO, Rome, 88 pp.

Gryseels, G. (1988) The role of livestock on mixed smallholder farms in the Ethiopian Highlands. Doctoral Dissertation University of Wageningen, 249 pp.

Hurni, H. (1998) Lessons of soil conservation research conducted in Ethiopia. *Proceedings Soil Fertility Management Workshop,* NFIA, Addis Ababa, Ethiopia, p. 31.

Jewitt, T.N., Law, R.D. and Virgo, K.J. (1979) Vertisol soils of the tropics and subtropics: their management and use. *Outlook on Agriculture* 10(1), 33–40.

Jutzi, S., Anderson, F. and Abiye Astake (1987) Low-cost modifications of the traditional Ethiopian tine plough for land shaping and surface drainage of heavy clay soil: Preliminary results from on-farm verification trials. *ILCA Bulletin* 27, 28–31.

Jutzi, S., Haque, I., McIntire, J. and Stares, J. (1988) Management of Vertisols in sub-Saharan Africa. In: Jutzi, S., Haque, I., McIntire, J. and Stares, J. (eds) *Proceedings of a Conference held at ILCA,* Addis Ababa, Ethiopia, 31 August to 4 September 1987, 431 pp.

Leather, J.W. (1898) On the composition of Indian soils. *Agricultural Ledger (Agriculture Series)* 24, 5(2).

Mulonga, W., Deckers, J. and Feyen, J. (1997) Changes in soil profile characteristics through cultivation of an upland Vertisol in Zambia. *Soil Use and Management* 13, 218–224.

Nachtergaele, T.O. (1976) Suleimi Series Benchmark soil description. *Technical Bulletin* No. 26, FAO/Soil Survey Administration, Wad Medani, Sudan.

Nyssen, J., Mitiku Haile, Moeyersons, J., Poesen, J. and Deckers, J. (1999) Soils and water conservation in Tigray (Northern Ethiopia): the traditional *daget* technique and its integration with introduced techniques. *Land Degradation and Development* 11, 199–208.

Sigunga, O.D. (1997) Fertilizer nitrogen use efficiency and nutrient uptake by maize (*Zea mays* L.) in Vertisols in Kenya. PhD Dissertation, Wageningen University, The Netherlands, 300 pp.

Spaargaren, O. (1994) *World Reference Base for Soil Resources*. Draft, ISSS–ISRIC–FAO, Rome, 161 pp.

Van Wambeke, A. (1991) *Soils of the Tropics. Properties and Appraisal*. McGraw-Hill, Ithaca, USA, 343 pp.

Soil and Water Conservation Strategies for Vertisols: Past Experiences and Challenges Ahead for Africa

2

J. HUSSEIN[1] AND M. ADEY[2]

[1]Department of Soil Science and Agricultural Engineering, University of Zimbabwe, Harare, Zimbabwe; [2]Department of Agriculture and Environmental Science, University of Newcastle upon Tyne, UK

Key words: Africa, Vertisols, soil and water conservation, soil erosion, semi-arid zones, Vetiveria, humid zones, pastures

Introduction

Vertisols occupy about 320 million ha worldwide; about 60% of this area is located in the tropics, 30% in the subtropics, and 10% elsewhere (Dudal and Eswaran, 1988). They are found under a range of climatic conditions, with 13% found in humid and sub-humid areas, 65% in semi-arid areas, 18% in arid areas and 4% under Mediterranean conditions. About 35% of the world's Vertisols occur in Africa (Ahmad, 1996a), mainly in semi-arid areas, although some countries, such as Sudan, have Vertisols straddling climatic zones ranging from humid to arid (Kanwar and Virmani, 1987). Vertisols, with their extremes in physical and chemical behaviour, require specific types of soil and water management strategies in each of these zones, suited not only to the climate, but also to the type of farming or natural resource use, the terrain and the prevailing socio-economic conditions. These strategies should maximize the unusual attributes of Vertisols to enhance each specific environment. For example, the large water storage potential of Vertisols in semi-arid areas should be promoted by improving the self-mulching properties of the soils, which in turn promotes greater infiltration and deep wetting of the profile (Hussein and Adey, 1994).

There has been a vast amount of research and development work on soil and water conservation (SWC) for Vertisols. In this chapter, it is not our intention to try and itemize this work, but rather to focus on the major progress achieved in different countries in controlling erosion and managing water resources in these unusual soils in the tropics and subtropics. Progress in identifying and using biological conservation techniques is also discussed. Additionally, it is very important to evaluate the economic costs and benefits of the different programmes, as those without clear benefits often fail to stand the test of time. We also wish to highlight why SWC programmes sometimes fail and what lessons have been learnt from these failures.

It is no good introducing measures to improve soil fertility, if there is going to be loss of nutrients, organic matter and soil through erosion. Any planned improvements in fertility must go hand in hand with improved SWC. In addition, it is worth emphasizing that improvements in SWC often act synergistically with improvements in other agronomic practices, to improve crop yields greatly (Kanwar and Virmani, 1987).

As Africa faces increasing economic and political problems, we hope to show that well-designed SWC programmes have made a difference to people's lives here and elsewhere and can contribute to the overall development of our continent.

The Erosion Problem

Severe erosion, particularly on cultivated Vertisols, is widespread, especially on land with slopes of 3% or more (Probert *et al.*, 1987). Water erosion is considered more of a problem than wind erosion on Vertisols, as the surfaces of these soils are frequently cloddy and rough, limiting susceptibility to wind erosion (Briggs, 1972). This chapter will therefore focus on water erosion.

Vertisols have slow infiltration when their cracks are closed and their inherent structural instability under rainfall leads to the formation of surface seals and crusts (Mullins *et al.*, 1987). Although many Vertisols are well structured, they slake under rapid wetting to form micro-aggregates in the fine sand to silt size range that are easily transported by water because of their low density (Loch and Donnollan, 1983). These factors result in high runoff and erosion rates under rainfall (Hussein *et al.*, 1992) and sometimes under irrigation (Loveday, 1984). This problem is worsened when the Vertisols occur on steep slopes. Management practices that exacerbate erosion include long fallow periods, frequent aggressive tillage, removal of crop residues and cropping of shallow soils (Mullins *et al.*, 1987). Soil movements of as high as 61 t ha^{-1} $year^{-1}$ have been reported in small catchments under bare summer fallow systems in Australia (Freebairn and Wockner, 1986).

Erosion includes sheet wash, rill erosion and gullying, with rill erosion being more important on cropped soils and sheet erosion on grazed Vertisols (Mullins *et al.*, 1987). Gullies, once initiated, can develop very rapidly due to the slumping nature of wet Vertisols. Briggs (1972) noted that, in Australia, gullies are frequently rectangular in cross-section and formed in a series of steps.

The general consequences of erosion include:

- loss of fertile topsoil with organic matter;
- loss of nutrients;
- reduced depth of soil, therefore less volume of soil to supply nutrients, water and anchorage;
- exposure of unfavourable subsoils, e.g. with higher sodium levels;
- gullies interfering with tillage, weeding, spraying and harvesting;
- soils deposited elsewhere leading to siltation of rivers, dams and roads;
- excessive runoff causing flooding; and
- agrochemicals washed off and deposited elsewhere causing eutrophication, pollution and poisoning.

Vertisols are prone to all of these problems. However, comparing Vertisols with other soils, the consequences of erosion may differ in some instances. For example, while loss of organic matter (OM) in the surface soil by erosion may be detrimental to the surface structure and nutrient supply, it is unlikely to reduce the cation exchange and water-holding capacities of Vertisols, because of their large active clay fractions. This is in contrast to sandy soils, where loss of OM may greatly diminish the nutrient and water retaining ability of the soil. In addition, reduction of OM with depth in Vertisols is typically not as marked as in non-Vertisolic soils, due to pedoturbation. Vertisols may therefore have reserves of OM in the sub-surface horizons. This is illustrated in Fig. 2.1, which compares variation in OM with depth for several soil types.

Exposure of unfavourable subsoils through erosion may also be a problem in some Vertisols that have increasing sodium content with depth (Table 2.1). Erosion of the topsoil may effectively move the sodic horizons nearer the surface, causing unfavourable physical and chemical characteristics and decreasing crop yield. Cook *et al.* (1992) noted that there was great similarity between the 0–10 cm layer of a cultivated Vertisol and the 10–20 cm layer of virgin soil, consistent with loss of the top 10 cm by erosion in the cultivated soil.

Soil and Water Conservation Technologies in Different Climatic Zones

Emphasis on soil and/or water conservation is different in the various climatic zones of the world. For example, soil conservation allied with safe removal of water may be important in humid zones while reduction in runoff and storage of water *in situ* could be critical in arid zones. The various types of conservation technologies have been fairly well established through many years of research. Such technologies include:

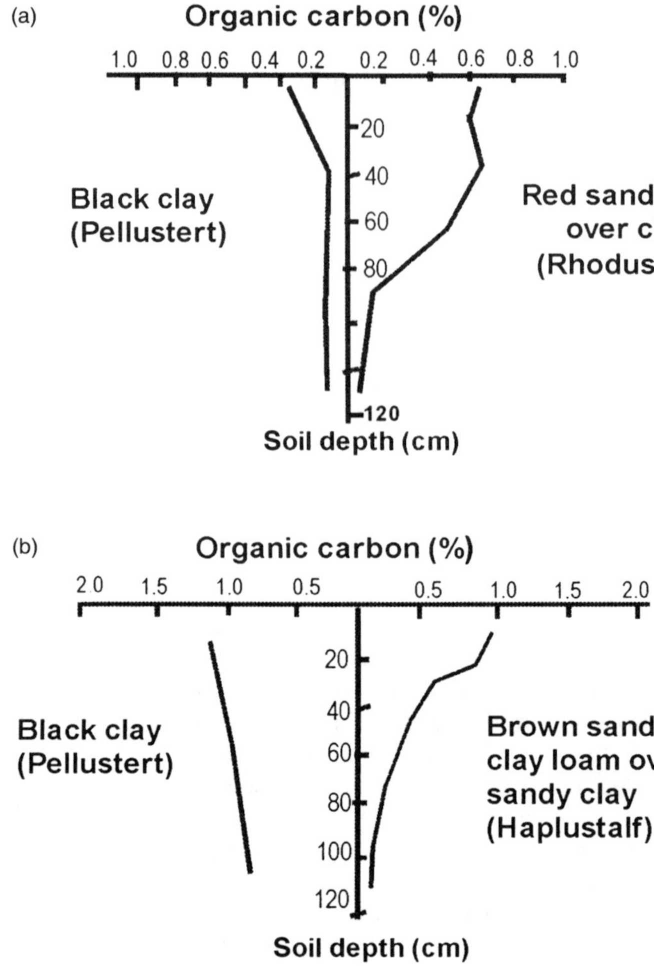

Fig. 2.1. Variation of organic matter with depth for paired soil types in different areas. (a) Soil profiles at ICRISAT Centre, Hyderabad, India. (Source: El-Swaify *et al.*, 1985.) (b) Soil profiles in Lowveld, Zimbabwe. (Source: Hussein, 1981.)

- mechanical conservation works, e.g. terraces, contour ridges, storm drains, channels, tied furrows;
- biological conservation, e.g. tree and cover crops, strip cropping, controlled grazing, surface mulching; and
- combinations of mechanical and biological conservation, e.g. grassed contour ridges, conservation tillage systems, grassed waterways.

Mechanical conservation technologies usually work by acting as physical barriers that slow down the velocity of runoff, while biological methods reduce raindrop impact and check runoff and soil loss (Young, 1997). Combination

Table 2.1. Some examples of Vertisol profiles with increasing exchangeable sodium percentage in the subsoil.

El-Swaify *et al.* (1985) Hyderabad, India		Sehgal *et al.* (1987) Achmati, India		Loveday (1984) Kerang, Australia	
Soil depth (cm)	Exchangeable sodium (%)	Soil depth (cm)	Exchangeable sodium (%)	Soil depth (cm)	Exchangeable sodium (%)
0–16	11	0–4	4	0–10	12
16–57	15	4–22	7	10–20	18
57–118	13	22–54	14	30–40	23
118–151	17	54–87	17	70–80	30

methods combine both approaches in varying degrees. There are many permutations of these technologies, appropriate for the many combinations of land use, e.g. grazing versus cropped areas, steep slopes versus flat areas, irrigated versus rainfed cropping, large-scale versus small-scale farmers.

Young (1997) notes that there has been a general change in conservation policy over time. Early approaches to soil conservation were focused primarily on the reducing the rates of soil loss with the use of primary earth structures such as contour-aligned banks, ditches, storm drains, and waterways. Extension work was often based on a top-down, legally enforced policy, either by prohibiting cultivation of steep slopes and waterways or with compulsory requirements for conservation work. Newer approaches to conservation, commonly called land husbandry or conservation farming, focus attention on the effects of crop cover and mulches in reducing soil loss. Thus, less emphasis is based on earth structures with more attention being given to biological conservation methods. In dry areas, greater emphasis is also being placed on integration of soil *and* water conservation.

Some of these conservation technologies for Vertisols are now discussed in the context of the various climatic zones, as defined by Troll's (1965) vegetation zone delineation.

Humid tropics

The humid tropics are those areas within the tropics where precipitation (P) exceeds potential evapotranspiration (PET) for more than 7 months of the year (Troll, 1965). In these areas, safe removal of excess water is usually a priority, together with reduction of high soil erosion rates. Vertisols are also frequently found in flat or low-lying areas within this zone, which compounds the problems of wetness and poor aeration. There are serious constraints on tillage operations as the optimum moisture content of the soil for tillage may only occur within very short periods.

Removal of excess water is best achieved by using appropriate forms of land layout such as ridges, broad banks, narrow beds and cambered beds on which major drainage installations (often using mechanical pumps) are superimposed (Ahmad, 1988). The choice of an appropriate crop is also crucial as crop growth usually occurs in near-saturated conditions. The influence of internal drainage in providing aeration can largely be ignored and emphasis has to be placed on adequate external drainage to remove excess precipitation. The external drainage layout must be designed and laid out carefully, to control the rate of runoff. Descriptions of different land layouts are shown in Table 2.2.

Semi-arid tropics

The semi-arid tropics (SAT) are those areas within the tropics where P exceeds PET for 2–7 months (Troll, 1965). This covers a broad range of rainfall zones and is further divided into wet SAT (P > PET for 4.5–7 months) and dry SAT (P > PET for 2–4.5 months). In the wet SAT, some areas may receive excessive rains during a short cropping season and safe removal and sometimes storage of this water is often a priority as well as the control of soil erosion. By contrast, in the dry SAT areas, the main constraint to cropping/grazing is usually low and unreliable rainfall during the cropping season and priority is usually given to water harvesting that will allow sufficient moisture to be trapped *in situ* to grow a crop. These two zones are discussed further.

Table 2.2. Different types of land layout used in humid areas. (Source: Ahmad, 1988.)

Land layout	Description of layout and crops grown	Constructed by:
Narrow ridges	About 0.5 m wide and 1 m apart made by ridging ploughs, e.g. used for sweet potato	Hand
Broad banks or ridges	About 2 m wide, e.g. used for vegetable crops	Hand/tractor
Narrow raised beds	About 4 m wide with crops planted on ridges made across or along the bed, e.g. used for field or vegetable crops	Tractor
Cambered beds	About 20 m wide. Crops planted on ridges across or along the bed. Provide good drainage and less soil erosion but crops may perform differentially from crest to sides of bed in response to soil fertility/drainage conditions, e.g. used for sugarcane	Bulldozer

Wet SAT

In the wet SAT in India, cropping on Vertisols has proved difficult in many areas. Many farmers used to leave these soils fallow during the monsoon and then crop after the monsoon, using residual moisture. This was due to problems of waterlogging and difficulties with tillage when the soil was at its wettest. Hydrological studies of this cropping system showed that rain falling during the monsoon was not used effectively with 28% lost as runoff with allied soil erosion (6.9 t ha^{-1} $year^{-1}$), 24% evaporated from the bare fallow, 9% lost as deep percolation and only 39% was used for post-rainy season cropping (El-Swaify *et al.*, 1985).

In response to this problem, the International Crops Research Institute for the Semi-Arid Tropics (ICRISAT) developed a broadbed and furrow (BBF) system that has proved very successful in controlling water and soil loss in India under rainfall of 750–1250 mm $year^{-1}$ (both semi-humid and semi-arid areas). The BBF consists of individually graded wide beds separated by furrows that drain into grassed waterways laid out at a gradient of 0.4–0.8% (Fig. 2.2). This system allows double cropping with one crop grown during the monsoon and another afterwards, using residual moisture. The BBF design is flexible and allows different crops to be grown in the beds by altering crop spacing, e.g. for

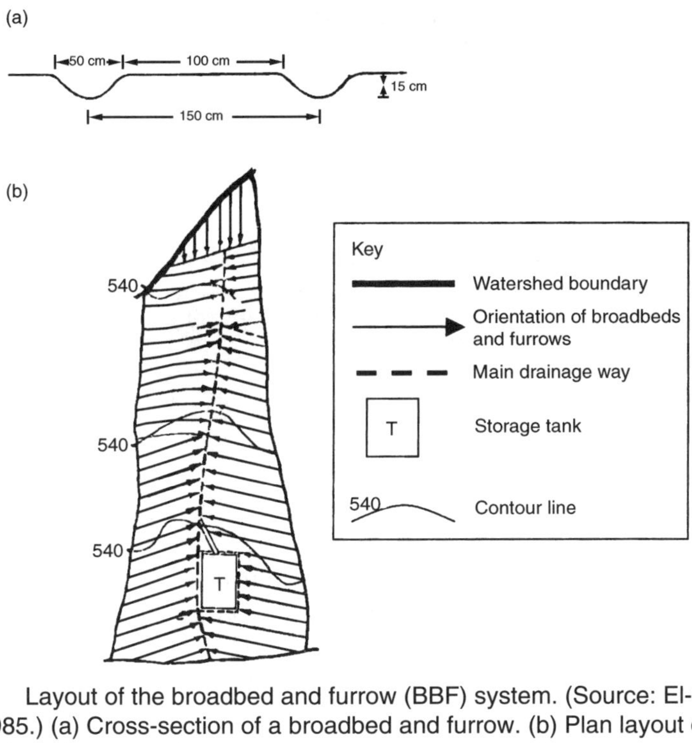

Fig. 2.2. Layout of the broadbed and furrow (BBF) system. (Source: El-Swaify *et al.*, 1985.) (a) Cross-section of a broadbed and furrow. (b) Plan layout of broadbed and furrow system.

maize two rows are sown, 75 cm apart, whereas for sorghum, three rows are used, 45 cm apart. Greatly increased yields have been achieved using this system (El-Swaify *et al.*, 1985).

The effectiveness of the BBF system in controlling soil loss and reducing runoff is shown in Table 2.3, which compares the BBF to the traditional flat monsoon fallow system. Runoff is reduced by half and soil loss to about one-sixth of the traditional value, using the BBF system.

The BBF technique has also been adapted to Ethiopian conditions. Traditionally, in the Ethiopian Highlands, cultivation almost exclusively uses animal power and crops are grown on a ridge/furrow system, 30 cm wide. Before the 1980s, beds were made rarely as they required too much labour. However the BBF technique allows better drainage, earlier planting and double cropping. The International Livestock Centre for Africa (ILCA) scientists and their collaborators have, therefore, designed a low-cost implement, the ILCA broadbed maker (BBM), which forms broad beds about 20 cm high and 1.2 m apart. This implement is made from two unmodified traditional Ethiopian ploughs, *mareshas*, with the addition of metal wings and a chain linking the two ploughs. Earlier models of the BBM proved expensive and difficult to procure, so after extensive trials and testing the BBM was further developed with a triangular design. By using the *mareshas* as the basis for the BBM, farmers do not need to invest in a completely new implement to form the broadbed. Jutzi *et al.* (1987) also note that there are many advantages to using unmodified *mareshas* as these are widely available throughout the country and farmers are already familiar with them.

Average crop yields using the BBM have been improved by as much as 330% (faba beans) and 130% (local durum wheat) with greatly increased

Table 2.3. Annual rainfall, runoff and soil loss for a cropped Vertisol in India using a broadbed and furrow (BBF) system and a traditional monsoon fallow system. (Source: Pathak *et al.*, 1983, quoted in El-Swaify *et al.*, 1985.)

		BBF at 0.6% slope		Traditional flat monsoon fallow	
Year	Annual rainfall (mm)	Runoff as % of seasonal rainfall	Annual soil loss (t ha^{-1})	Runoff as % of seasonal rainfall	Annual soil loss (t ha^{-1})
1975	1041	15.6	1.39	24.0	5.21
1976	687	10.6	0.98	33.3	9.20
1977	585	0.2	0.07	9.0	1.68
1978	1125	24.3	2.93	36.7	9.68
1979	690	10.6	0.70	29.6	9.47
1980	730	15.9	0.97	24.1	4.58
Mean (1975–1980)	810	12.9	1.17	26.1	6.64

gross revenues and return on labour (International Livestock Centre for Africa, 1988). However, different cropping options need further research (Astatke and Mahomed-Saleem, 1998) to improve the water use efficiency of the system.

The *maresha* is also used as the basis for a terracing plough (Jutzi *et al.*, 1987) used to make 4-m-wide level terraces on slopes of 8%.

Dry SAT

In the dry SAT, tied furrows (TF), also called tied ridges, have been used successfully in Zimbabwe and elsewhere to reduce runoff and increase yields. The tied furrows trap water *in situ* and encourage infiltration and water storage. Vertisols are ideal soils for storing water as they hold large amounts of plant-available water (Hussein and Adey, 1994) if infiltration is facilitated, provided they are not sodic, when the effective available water capacity can be reduced drastically (Kutílek, 1983). Nyamudeza *et al.* (1992) reported that, on Vertisols, TF increased yields of sorghum, cotton and maize by 25, 32 and 46%, respectively, and gave higher gross margins in trials from 1982 to 1989. The crops were planted in the furrows with the 1.5 m TF generally outperforming the 1.0 m TF. However, crop rotations introduced some variability into the system and crops grown after cotton and sorghum in dry years experienced moisture stresses due to the reduction in stored soil moisture. In very wet years, crops do not grow well on TF due to excessive wetness; Mitchell (1987) found that in wet years, in Malawi, cotton yields were decreased when grown on tied ridges compared with levelled land. The improved soil moisture conservation under the TF system is illustrated in Fig. 2.3.

No data are available for soil losses under this system on Zimbabwean Vertisols although similar trials on sandy and red clay soils under no-till, tied ridging (NTTR) showed considerably reduced soil loss and runoff compared with flat cropping (Chuma and Hagmann, 1997).

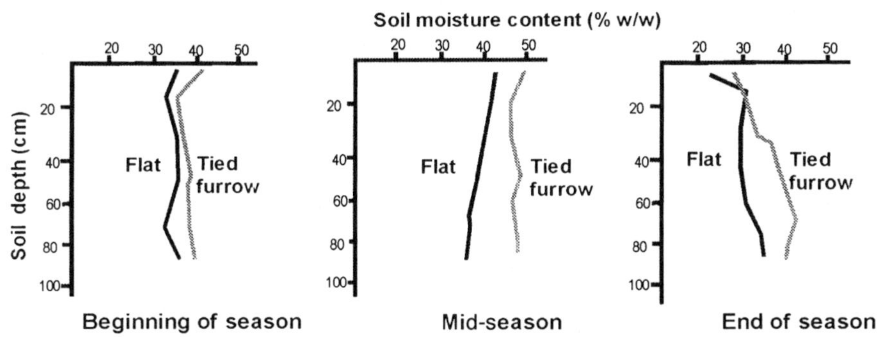

Fig. 2.3. The effect of land shaping on soil moisture storage under sorghum in the 1989/90 season with total rainfall of 545 mm. (Source: Nyamudeza *et al.*, 1991.)

Arid tropics

In Africa, Vertisols in parts of Chad, Somalia and Sudan occur in arid zones where P > PET for less than 2 months a year (Kanwar and Virmani, 1987). Under such dry conditions, rainfed cropping is severely constrained and Vertisols are more likely to be used for grazing purposes only. Systems such as TF may be successful in good years with short-season crops, but generally moisture is too limiting for successful cropping, unless irrigation is available.

SWC Through Land Planning and Conservation Farming

The importance of overall land planning for SWC is stressed by Yule (1987), who notes that a wide range of options exist to manage the soil surface, so as to manipulate the soil water balance. Young (1997) suggests that management systems should be developed to incorporate these options. These systems should be based on zonal tillage, i.e. managing the zones of the field for their intended use with zones for traffic, infiltration, root growth, etc. These systems are usually designed around permanent beds and thus provide excellent control of machine operations. Examples include the BBF and TF systems.

The theme of land-use planning and land husbandry in Australia is extended in an article by Mullins *et al.* (1987), who note that, in order to conserve soil and water, there is a need to: maximize infiltration, protect the soil surface against raindrop impact and wind erosion and control runoff. Protection of the soil surface can be achieved by using high stubble-producing crops such as wheat/barley and by reducing the proportion of low crop-cover crops such as cotton or sunflower in the system. Tillage operations should be reduced or controlled, e.g. by use of herbicides instead of mechanical weeding. Crop rotations that only include short fallows should be used.

A watershed approach to management of soil and water resources has gained momentum in recent years. This approach extends SWC from a plot or field scale to a watershed/catchment scale, by examining the overall water flow and soil movement within the catchment. A more broadly based and efficient management system can then be introduced, instead of merely implementing changes on a plot or field by field basis. Vittal *et al.* (1996) examined SWC using one of the 47 model watersheds developed by the Government of India. The programme consisted of measures such as percolation tanks, grading checks and deep tillage, on which were superimposed improved crop management practices such as improved cultivars, addition of fertilizers and timeliness of operations. Runoff was reduced by 26%, ground cover increased by 38%, due to increased productivity, and income per capita increased by 67%. By increasing infiltration and reducing runoff, water-table levels were raised by 50 cm, allowing sugarcane to be introduced.

Correct layout and incorporation of SWC measures have been effective in preventing erosion on newly opened wheat farms in Tanzania (Ngatoluwa *et al.*, 1987), whereas older farms were subject to severe erosion as conservation measures were not in place. Conservation measures on new farms include contoured grass strips (5–8 m apart) and channel terraces (shallow, wide ditches) installed on the contour (75–125 m apart on slopes of > 3%). These intercept and store water. On steeper slopes with shallow soils, there are also graded channel terraces with a gentle slope of 0.3%, to conduct water to a waterway.

In a study of soil erosion on Vertisols of the Eastern Darling Downs in Australia, Freebairn and Wockner (1986) noted that the area was susceptible to high erosion rates due to the combination of erodible clays, steep slopes up to 10%, high-intensity summer rain and cropping practices that left soils with little protective cover during summer. Even though 60% of this area had some form of conservation structures such as contour banks and grassed waterways, erosion was still prevalent. This area grows winter wheat but frequently there is no summer crop to shield the soil from high-intensity summer rainfall. Conservation tillage practices have been developed (Unger and Stewart, 1988), which result in greater proportions of stubble being left on the soil surface. The reduced cultivations under such systems also result in more stable macro-pores, which retain their integrity even as the soil wets and swells (Turpin *et al.*, 1999). Freebairn and Wockner (1986) compared runoff and soil movement over 8 years, under five different summer management systems: bare fallow, stubble incorporated, stubble mulch, zero tillage and a summer crop (sorghum, sunflower or maize) (Table 2.4). They found that runoff and sediment load were related inversely to surface cover although the relationships differed depending on the antecedent soil moisture. When the profile was very wet, runoff rates were very high even under good cover but sediment loads were reduced due to reduced runoff velocity and decreased raindrop

Table 2.4. Summary of the effect of different types of summer cover on annual runoff and soil movement during 1978–1984 for two Vertisols (a black earth at Greenmount and a grey clay at Greenwood) on Eastern Darling Downs, Australia. (Source: Freebairn and Wockner, 1986.)

Summer cover	Bare fallow	Stubble incorporated	Stubble mulch	Zero tillage	Summer crop
Greenmount site					
Mean annual runoff as % of mean rainfall	11	8	7	9	7
Mean annual soil movement (t ha^{-1})	61	17.9	5.3	2.1	22.3
Greenwood site					
Mean annual runoff as % of mean rainfall	11	9	8	8	8
Mean annual soil movement (t ha^{-1})	31.6	7.8	3.9	1.8	19.8

impact. Zero tillage was the most effective in reducing soil erosion and bare fallow the least effective. Freebairn and Wockner (1986) concluded that soil movement could be reduced by 80–90% if a soil cover of 20–30% were maintained during the most erosive period (October–March).

Working in the Sudan, Willcocks (1984) compared different tillage techniques on Vertisols. He found that the energy cost per unit yield for manual, no-tillage planting was about 25% of that of more vigorous chisel ploughing and was quicker, which allowed more timely and extensive seedbed preparation.

Pastures

Where Vertisols occur in regions of low rainfall, land use may be restricted to grazing, as often found in Africa and Australia (Probert *et al.*, 1987). In many areas, overgrazing is a problem, leaving little cover on the land. This results in serious soil loss and runoff from these areas. While it is technically feasible to improve pastures on Vertisols, the costs may be prohibitive. Additionally, there is often little incentive to invest in pasture improvement under traditional systems of communal grazing. Research in Australia has had little success in finding legumes suitable for improving pastures on drier Vertisols in summer rainfall areas (Russell, 1984) as plant establishment is frequently poor. Rhodes grass (*Chloris gayana*) and green panic (*Panicum maximum* var.) have low survival rates after seeding, mainly due to water stresses, mechanical impedance to seedling germination, and high temperatures during summer months. Salinity at depth also contributes towards the poor growth of legumes. Pre-germination of seeds and water injection were suggested as ways of improving plant establishment. More research in this area is required to identify not only suitable pasture species, but also techniques to establish and maintain good cover in pastures in semi-arid Vertisols. For the wetter Vertisols, there are fewer problems in establishment and survival, with a range of grasses available for improving pastures including Rhodes grass and green panic as well as buffel (*Cenchrus ciliaris*) and pangola grass (*Digitaria decumbens*) (Probert *et al.*, 1987).

Developments in Biological Conservation

There has been a considerable amount of development in the field of biological conservation since the 1980s. This has focused on the use of vegetative barriers, such as vetiver grass (Truong, 1998), agroforestry for soil conservation and management (Young, 1997) and conservation tillage systems (as discussed in the previous section).

Vetiver (*Vetiveria zizanoides*) is well suited to Vertisols as it can withstand the extremes of moisture and temperature associated with these soils. Certain varieties such as Monto vetiver are also tolerant of the saline and sodic conditions frequently found in Vertisols (Truong, 1998). Sterile varieties are usually recommended for SWC programmes. Vetiver hedges are more effective than leucaena hedges and lemon grass at some sites in India, in reducing runoff and soil loss and increasing grain yield (Truong, 1998). Vetiver and leucaena also increased available moisture more than *Cymbopogon flexuosus* and *Chrysopogon martinii* according to Rane *et al.* (1995) in Maharashra, India. Similarly, vetiver hedges have been very effective in controlling floods in the Vertisol areas in Australia (Knowles-Jackson, 1998) and were as effective in controlling seasonal runoff and soil loss as mechanical barriers in Madhya Pradesh, India (Ranade *et al.*, 1997). Vetiver has also been used to stabilize gullies when planted on the contour line above the gully head.

During the last decade, there has been a great increase in the identification and promotion of multi-purpose trees that can be used for SWC on Vertisols, particularly in India. However, much more research and development is required to improve the availability, management and use of suitable trees. This is not easy, as tree growth can be very difficult on Vertisols. It is thought that the damage to roots during seasonal cracking of Vertisols limits the number of tree species that can grow on Vertisols. Frequency of fire is also a likely factor reducing natural tree establishment (Probert *et al.*, 1987). However, trees can be very useful in SWC programmes. They are used to reduce erosion by increasing soil cover (litter), acting as a runoff barrier (e.g. closely planted hedgerows), to strengthen and stabilize earth structures through the binding activity of roots, and to provide benefits (fuel, fruit, fodder) on otherwise unused earth conservation structures.

Acacia, *Eucalyptus* and *Casuarina* species can tolerate seasonally variable conditions and moderate levels of salinity and, therefore, grow tolerably well on Australian Vertisols (Probert *et al.*, 1987). Hebbara *et al.* (1992) reported that, from 23 trees screened for growth on a saline Vertisol in India, *Casuarina equisetifolia*, *Gliricidia sepium*, *Inga dulcis*, *Acacia auriculiformis*, *Dalbergia sissoo* and *Acacia nilotica* performed the best. *Leucaena leucocephala* hedgerows have also been grown for fodder production on Vertisols in India (Korwar, 1998) and when used as mulch (loppings) increased the infiltration rate and water retention (Bellakki and Badanur, 1994). Nagpur mandarin and *Ziziphus mauritiana* have been grown on Vertisols for fruit in India (Itnal *et al.*, 1993) and *Eucalyptus* trees are grown for fuelwood on Vertisols at Chisumbanje, Zimbabwe. Coolibah (*Eucalyptus microtheca*) has been successful in woodlots in the Gezira, Sudan.

Returns on Soil and Water Conservation Investment

While Enters (2000) has summarized effectively the challenges to consistently judging the economic benefit of soil conservation, it is notoriously difficult to assess the returns to conservation investment (Scoones *et al.*, 1996). There are few such studies in Africa. Too often SWC is assumed to be automatically beneficial without looking in detail at the costs and benefits. Scoones *et al.* (1996) report that the few studies that have been carried out relate to large-scale mechanical conservation works and show in most cases that returns are negative. Ryan and Sarin (1981, quoted in Ahmad, 1996b) reported economic returns on smaller-scale investments such as BBF in India.

Some proponents of SWC programmes argue that calculation of immediate returns should not be a concern since measures are aimed at long-term conservation, which must attract external subsidy in order to ensure intergenerational equity and to offset the wider costs of erosion. However, cost effectiveness, even in the short term, will remain a priority for farmers and project planners – experience has shown that while subsidies may well alter behaviour and encourage investment in SWC, as soon as the subsidy is withdrawn farmers switch to cost-effective strategies.

Investment in SWC depends on the willingness of farmers to expend labour for increased benefits that may be obtained for instance in the first year from water harvesting in semi-arid regions or, alternatively, some time in the future in agroforestry practices. This means people must feel confident of secure benefits from their investment. However insecurity can arise from problems such as conflict and war and insecurity of land tenure.

The importance of economic viability is again stressed by Tisdell (1996). Multiple criteria for sustainability, such as the five pillars in the International Board for Soil Research and Management's (IBSRAM, 1991) Framework for the Evaluation of Sustainable Land Management (FESLM), raise difficult assessment and evaluation problems. Tisdell (1996) notes that economic viability is included as one of the five pillars of FESLM and is essential for the sustained adoption of conservation farming projects. However, the economic viability is difficult to assess as it depends on a variety of attributes in the farming system and a holistic approach is needed in evaluating it.

Adoption of Soil and Water Conservation Technologies

Whilst many of the technologies for SWC have been known for some time, adoption by farmers has not been as widespread as hoped. This is due to a number of socio-economic and technical factors (Hagmann *et al.*, 1996a,b), including:

- a heavy-handed, top-down approach to extension;
- political connotations attached to certain conservation structures;
- insensitivity of extensionists to problems facing women farmers;
- inappropriateness of some mechanical conservation works to the environmental conditions; and
- limitation of conservation measures to a 'one type fits all' scenario.

In Africa, another major constraint to the adoption of some land-farming practices has been the availability of power equipment (FAO, 1987). Some of the SWC technologies, such as cambered beds and TF, require substantial soil movement, which is facilitated by the availability of tractors. However, once formed, many of these conservation structures can be maintained by hand.

In response to these constraints, changes in extension approaches have now been made in many cases, to enable the participatory development of flexible options for SWC. These can be modified and adapted to the specific environment and optimized by the farmers themselves. However, participatory approaches, while clearly desirable, are not simple. Nevertheless, it is better to encourage a process of negotiation and choice involving all parties in development planning and implementation. Technology exists therefore not simply as an engineering design but in a social and economic context. So for technology to be attuned to people's needs, local environmental conditions and economic factors, it must be flexible and adaptable. Rigid prescriptions and designs do not work. Indigenous technologies should also be synthesized with proven techniques and adapted to specific sites, situations and farmers' needs (Hagmann and Murwira, 1996).

Hudson (1991), in his review of the reasons for success or failure of soil conservation projects, similarly notes that farmer involvement should start at the time of project preparation and not at its implementation. Hudson further stresses that the technology used for the project should be easy to teach and demonstrate and for widespread adoption there must be a multiplier effect of farmers teaching other farmers.

A holistic approach to SWC is advocated by Young (1987), who states 'There should be no soil conservation projects. Conservation should be applied as an integral part of agricultural improvements linked to increases in productivity. In extension of SWC, farmers must be able to see tangible benefits in the form of higher production and associate these with conservation.' This re-emphasizes the points made in the previous section relating to the returns on SWC.

In one of the few studies specifically reporting about the adoption of SWC for Vertisols, Nyamudeza *et al.* (1992) examined the TF technique on Vertisols in the lowveld of Zimbabwe. An informal survey done after five seasons of on-farm trials indicated that 50–60% of the farmers were satisfied by the yield-improving effect of the TF technique. A later formal survey showed that a good number of farmers had adopted the technique and 93% of these found that it increased crop yields (i.e. they realized a tangible return to SWC).

It is hoped that, as researchers, planners and extensionists use these newer and more flexible participatory approaches, there will be an improved adoption rate of SWC measures, leading to an improvement in the productivity and sustainability of farming systems.

Research and Extension: Needs and Opportunities

As we consider all the points raised in our previous discussion, a number of ideas for further research and extension emerge:

1. Biological control of runoff and soil loss on Vertisols has gained momentum in recent years. However, there is still a long way to go and further research into biological conservation methods, especially identification and promotion of appropriate grass and multi-purpose tree species that grow well on Vertisols in semi-arid areas, should be considered as a priority in SWC research.
2. SWC strategies for Vertisols in Africa should be based on proven technologies within the different climatic zones but should be flexible enough to take into account variations in cropping practices, soil variations, slope and socio-economic conditions. Programmes that include both biological and mechanical conservation technologies should be encouraged as they are likely to be more successful, especially if the overall design is done on a catchment basis. Further information is required on the TF system, in terms of moisture storage, patterns of water extraction under different crops, and measurement of runoff and soil losses in comparison with flat planting.
3. Improvement of pastures on Vertisols is difficult as many of the traditionally used legumes do not grow well, particularly in drier regions. Further work on appropriate fodder species that grow well on Vertisols is required, particularly for the semi-arid areas. In addition, there should be some quantification of runoff and soil losses under different grazing systems so that informed decisions about the most appropriate layout and management of these systems can be made.
4. In many SWC programmes, the returns to conservation investment are difficult to assess. However, economic viability is considered crucial if these programmes are to succeed in the long term and conservation should be applied as an integral part of agricultural improvements linked to increases in productivity. More SWC studies, particularly in Africa, should therefore incorporate assessment of both short- and long-term economic viability.
5. More effort is required to improve the rate of adoption and adaption of SWC technologies on Vertisols. Researchers, planners and extensionists need to work in a participatory way with farmers, involving them from the very beginning in any new development or project. The short- and long-term returns to investment in SWC need to be as attractive as possible to the farmers, to encourage adoption. For widespread adoption of appropriate technologies, these should be simple and easy to demonstrate and spread from farmer to farmer.

Conclusions

Vertisols are very prone to runoff and erosion when wet, particularly where they occur on steep slopes and have poor vegetative cover and when subjected to intense rainfall. This has caused problems in farmlands in both developed and developing countries. Loss of surface OM through erosion is not likely to diminish the nutrient and water retention capacities of Vertisols, but other erosion problems such as structural instability, reduction in nutrient cycling, and exposure of sodic subsoils may be critical. It is no good improving soil fertility if erosion and runoff are not controlled. SWC often acts synergistically with improvements in other agronomic practices to greatly improve crop yield. The two must go hand in hand if benefits from improved soil fertility are to be realized.

In humid regions, safe removal of excess water by correctly laid out drainage systems is a priority. Crops grow best on raised beds as aeration is improved. In the wet SAT, the BBF system is effective in reducing runoff and soil loss and improving crop yields. In the dry SAT, *in situ* water conservation is important and techniques such as TF cropping are successful, except in very wet years.

SWC programmes should involve broad land-use planning and good land husbandry through technologies such as watershed management and conservation farming. Conservation tillage systems that leave good cover on the ground greatly reduce runoff and soil loss on Vertisols, with zero tillage usually being the most effective. Use of biological conservation techniques has gained momentum over the last decade but there is great scope for the expansion in their use.

It is hoped that, as researchers, planners, and extensionists use newer and more flexible participatory approaches to working with farmers, there will be a greater adoption rate of SWC measures, leading to an improvement in the productivity and sustainability of farming systems. This is particularly imperative in Africa, to reduce poverty and improve food security and the long-term sustainability of the environment.

References

Ahmad, N. (1988) Management of Vertisols in the humid tropics. In: Wilding, L.P. and Puentes, R. (eds) *Vertisols: Their Distribution, Properties, Classification and Management*. SMSS Technical Monograph No. 18, Texas A&M University, College Station, USA, pp. 97–115.

Ahmad, N. (1996a) Occurrence and distribution of Vertisols. In: Ahmad, N. and Mermut, A. (eds) *Vertisols and Technologies for their Management*. Developments in Soil Science No. 24, Elsevier, Amsterdam, The Netherlands, pp. 1–41.

Ahmad, N. (1996b) Management of Vertisols in rainfed conditions. In: Ahmad, N. and Mermut, A. (eds) *Vertisols and Technologies for their Management*. Developments in Soil Science No. 24, Elsevier, Amsterdam, The Netherlands, pp. 66–79.

Astatke, A. and Mahomed-Saleem, M.A. (1998) Effect of different cropping options on plant available water of surface drained Vertisols in the Ethiopian Highlands. *Agricultural Water Management* 36, 111–120.

Bellakki, M.A. and Badanur, V.P. (1994) Effect of crop residue incorporation on physical and chemical properties of a Vertisol and yield of sorghum. *Journal of the Indian Society of Soil Science* 42(4), 533–535.

Briggs, H.S. (1972) The erosion process and swelling clay soils. In: *Physical Aspects of Swelling Clay Soils*. A symposium held in February. University of New England, Armidale, Australia.

Chuma, E. and Hagmann, J. (1997) Conservation tillage techniques for semiarid Zimbabwe: Results and experiences from on-station and on-farm interactive innovation in Masvingo (1988–1994). *Zimbabwe Science News* 31, 34–41.

Cook, G.D., So, H.B. and Dalal, R.C. (1992) Structural degradation of two Vertisols under continuous cultivation. *Soil and Tillage Research* 24, 47–64.

Dudal, R. and Eswaran, H. (1988) Distribution, properties and classification of Vertisols. In: Wilding, L.P. and Puentes, R. (eds) *Vertisols: Their Distribution, Properties, Classification and Management*. SMSS Technical Monograph No. 18, Texas A&M University, College Station, USA, pp. 1–22.

El-Swaify, S.A., Pathak, P., Rego, T.J. and Singh, S. (1985) Soil management for optimized productivity under rainfed conditions in the semiarid tropics. In: Stewart, B.A. (ed.) *Advances in Soil Science*, Vol. 1. Springer-Verlag, New York, pp. 1–63.

Enters, T. (2000) *Methods for the Economic Assessment of the On- and Off-site Impacts of Soil Erosion*, 2nd edn. Issues in Sustainable Land Management No. 2. IBSRAM, Bangkok, Thailand.

FAO (1987) *African Agriculture: the Next 25 Years*. FAO, Rome.

Freebairn, D.M. and Wockner, G.H. (1986) A study of soil erosion of the Eastern Darling Downs, Queensland. I Effects of surface conditions on soil movement within contour bay catchments. *Australian Journal of Soil Research* 24, 135–158.

Hagmann, J. and Murwira, K. (1996) *Indigenous Soil and Water Conservation in Southern Zimbabwe: a Study on Techniques, Historical Changes and Recent Developments under Participatory Research and Extension*. Dryland Programme Paper No. 23, International Institute for Environment and Development, London, UK, 24 pp.

Hagmann, J., Chuma, E. and Gundani, O. (1996a) *Acknowledging the Role of Gender in Agricultural Research and Extension*. Report to GTZ, Section 04 Strategic Corporate Development, April 1996, London.

Hagmann, J., Murwira, K. and Chuma, E. (1996b) Learning together: development and extension of soil and water conservation in Zimbabwe. *Quarterly Journal of International Agriculture* 35, 142–162.

Hebbara, M., Viswanath, D.P. and Devarnavadagi, S.B. (1992) Performance of tree species on marginally saline Vertisols of Tungabhadra Project Area. *Journal of the Indian Society of Soil Science* 40(3), 616–617.

Hudson, N.W. (1991) A study of the reasons for success or failure of soil conservation projects. *FAO Soils Bulletin* No. 64. FAO, Rome.

Hussein, J. (1981) An investigation into soil-water relationships of some irrigated Vertisols derived from basalt in Zimbabwe. MSc thesis, University of Stellenbosch, South Africa.

Hussein, J. and Adey, M.A. (1994) Sustainable regeneration of surface tilth in Zimbabwean Vertisols through water management. In: Syers, J.K.S. and Rimmer, D.L. (eds) *Soil Science and Sustainable Land Management in the Tropics*. CAB International, Wallingford, UK, pp. 120–131.

Hussein, J., Adey, M.A. and Elwell, H.A. (1992) Irrigation and dryland cultivation effects on the surface properties and erodibility of a Zimbabwe Vertisol. *Soil Use and Management* 8, 97–103.

International Board for Soil Research and Management (IBSRAM) (1991) *Evaluation for Sustainable Land Management in the Developing World*, Vol. I. IBSRAM Proceedings No. 12, IBSRAM, Bangkok, Thailand, 81 pp.

International Livestock Centre for Africa (ILCA) (1988) Making technology more appropriate. *ILCA Newsletter* 7(2), 1–4 (April).

Itnal, C.J., Surakod, V.S. and Sajjan, G.C. (1993) Alternative landuse systems for black soils of Karnataka. *Rainfed Agricultural Research Newletter* (India) 2 & 3, 14–15.

Jutzi, S., Anderson, F.M. and Astatke, A. (1987) Low cost modifications of the traditional Ethiopian tine plough for land shaping and surface drainage of heavy clay soils: preliminary results from on-farm trials. *ILCA Bulletin* 27, 28–31 (April).

Kanwar, J.S. and Virmani, S.M. (1987) Management of Vertisols for improved crop production in the semiarid tropics: a plan for a technology transfer network in Africa. In: *Management of Vertisols under Semiarid Conditions*. IBSRAM Proceedings No. 6, IBSRAM, Bangkok, Thailand, pp. 157–172.

Knowles-Jackson, C. (1998) Application of vetiver grass in soil erosion and sediment control on the Darling Downs. Paper presented at a vetiver workshop, Towoomba, Australia, 6–7 November, 3 pp.

Korwar, G.R. (1998) Fodder production potential of leucaena hedgerows on an alfisol and a vertisol in the semiarid tropics. In: Gill, A.S., Roy, R.D. and Bajpal, C.K. (eds) *Nitrogen Fixing Trees for Fodder Production*. Proceedings of an International Workshop organized by FACTNET. Winrock International, Morrilton, USA, pp. 146–153.

Kutilek, M. (1983) Soil physical properties of saline and alkali vertisols. In: *Isotope and Radiation Techniques in Soil Physics and Irrigation Studies*. IAEA, Vienna, Austria, pp. 179–190.

Loch, R.J. and Donnollan, T.E. (1983) Field rainfall simulator studies on two clay soils of the Darling Downs, Queensland. II. Aggregate breakdown, sediment properties and soil erodibility. *Australian Journal of Soil Research* 21, 47–58.

Loveday, J. (1984) Management of Vertisols under irrigated agriculture. In: McGarity, J.M., Hoult, E.H. and So, H.B. (eds) *The Properties and Utilization of Cracking Clay Soils*. Reviews in Rural Science 5, University of New England, Armidale, pp. 269–277.

Mitchell, A.J.B. (1987) Management problems of cotton on Vertisols in the lower Shire Valley of Malawi. In: *Management of Vertisols under Semiarid Conditions*. IBSRAM Proceedings No. 6, IBSRAM, Bangkok, Thailand, pp. 221–229.

Mullins, J.A., Webb, A.A. and Coughlan, K.J. (1987) Approaches to erosion control on Vertisols in Queensland, Australia. In: *Management of Vertisols under Semiarid Conditions*. IBSRAM Proceedings No. 6, IBSRAM, Bangkok, Thailand, pp. 285–297.

Ngatoluwa, R., Mansoor, H., Majanga, B.D. and McKeague, J.A. (1987) Conservation measures in Vertisols under large scale wheat farming in northern Tanzania.. In: *Management of Vertisols under Semiarid Conditions*. IBSRAM Proceedings No. 6, IBSRAM, Bangkok, Thailand, pp. 145–154.

Nyamudeza, P., Mandiringana, O.T., Busanagavanye, T. and Jones, E. (1991) The development of sustainable management of Vertisols in the lowveld area of southeastern Zimbabwe. *IBSRAM Newsletter* No. 20 (June), 6–9.

Nyamudeza, P., Mazhangara, E. and Kamba, E. (1992) Adoption of the tied furrow technique and the effects of the technique and previous crop on residual moisture and yields of sorghum and maize. Paper presented to the Annual Review Meeting of the IBSRAM Vertisol Management Network in Africa, Accra Ghana, June 1992.

Pathak, P., Miranda, S.M. and El-Swaify, S.A. (1983) Improving rainfed farming for the semiarid tropics. Implications for soil and water conservation. In: *Proceedings of an International Conference on Soil Erosion and Conservation.* Honolulu, Hawaii, USA, January, 1983.

Probert, M.E., Fergus, I.F., McGarry, D., Thompson, C.H. and Russell, S. (1987) *The Properties and Management of Vertisols.* CAB International, Wallingford, UK.

Ranade, D.H., Jain, L.K., Gupta, R.K. and McCool, D.K. (1997) Comparative performance of mechanical and vegetative soil conservation measures in Vertisols. *Indian Journal of Soil Conservation* 25, 182–185.

Rane, P.V., Sagare, B.N. and Rewatkar, S.S. (1995) Available soil moisture storage and nutrient uptake by cotton as influenced by vegetative barriers. *Annals of Plant Physiology* 9(2), 139–141.

Russell, J.S. (1984) Management of cracking clay soils in rainfed environments. In: *Physical Aspects of Swelling Clay Soils.* A symposium held in February, University of New England, Armidale, Australia, pp. 307–315.

Ryan, J.G. and Sarin, R. (1981) Economics of technology options for Vertisols in the relatively dependable rainfall regions of the Indian semiarid tropics. In: *Improving the Management of India's Deep Black Soils.* Proceedings of seminar on management of Deep Black Soils for increased production of cereals, pulses and oilseeds, New Delhi, India, pp. 37–57.

Scoones, I., Reij, C. and Toulmin, C. (1996) *Sustaining the Soil: Indigenous Soil and Water Conservation in Africa.* Drylands Programme Paper No. 67, International Institute for Environment and Development, London, UK, December, 25 pp.

Sehgal, J.L., Vadivelu, S., Hirekerur, L.R. and Deshpande, S.B. (1987) Soil criteria for selecting experimental sites for a Vertisol management network. In: *Management of Vertisols under Semiarid Conditions.* IBSRAM Proceedings No. 6, IBSRAM, Bangkok, Thailand, pp. 71–82.

Tisdell, C. (1996) Economic indicators to assess the sustainability of conservation farming projects: an evaluation. *Agriculture, Ecosystems and Environment* 57, 117–131.

Troll, C. (1965) Seasonal climates of the Earth. In: Rodenwaldt, E. and Justatz, H. (eds) *World Maps of Climatology.* Springer Verlag, Berlin.

Truong, P. (1998) *An Overview of Research, Development and Application of the Vetiver Grass System (VGS) Overseas and in Queensland.* Vetiver Workshop, Toowomba, Australia.

Turpin, J.E., Bridge, B.J., Oraye, D. and Thompson, J.P. (1999) Water and bromide movement in a Vertisol under four management systems. *Australian Journal of Soil Research* 37, 75–81.

Unger, P.W. and Stewart, B.A. (1988) Conservation techniques for Vertisols. In: Wilding, L.P. and Puentes, R. (eds) *Vertisols: Their Distribution, Properties, Classification and Management.* Technical Monograph No. 18, Soil Management Support, Services, Texas A&M University, pp. 165–181.

Vittal, K.P.R., Das, S.K., Katyal, J.C., Munikrishnaiah, N. and Reddy, M.R. (1996) Soil and water conservation and improved crop management effects on watershed productivity in Andhra Pradesh, India. *American Journal of Alternative Agriculture* 11, 2–6.

Willcocks, T.J. (1984) Tillage requirements in relation to soil type in semiarid rainfed agriculture. *Journal of Agricultural Engineering Research* 30, 327–336.

Young, A. (1987) The potential of agroforestry for soil conservation, with special reference to Vertisols. In: *Management of Vertisols under Semiarid Conditions.* IBSRAM Proceedings No. 6, IBSRAM, Bangkok, Thailand, pp. 187–199.

Young, A. (1997) *Agroforestry for Soil Management.* CAB International, Wallingford, UK, in association with ICRAF.

Yule, D. (1987) Water management of Vertisols in the semiarid tropics. In: *Management of Vertisols under Semiarid Conditions.* IBSRAM Proceedings No. 6, IBSRAM, Bangkok, Thailand, pp. 107–123.

Sustainable Nutrient Management of Vertisols 3

J.K. SYERS,[1] P. NYAMUDEZA[2] AND Y. AHENKORAH[3]

[1]*Department of Natural Resources and Environmental Sciences, Naresuan University, Phitsanulok, Thailand;* [2]*Department of Research and Specialist Services, Chiredzi Research Station, Chiredzi, Zimbabwe;* [3]*Department of Soil Science, University of Ghana, Legon, Ghana*

Key words: Vertisols, nutrients, organic carbon, nitrogen, phosphorus, potassium

Introduction

Vertisols are regarded as being inherently fertile soils but such a generalization can mask quite a wide range of chemical fertility characteristics. For example, whereas the levels of total and exchangeable potassium (K) are usually high and thus deficiences in crops rare (Finck and Venkateswarlu, 1982), nitrogen (N) deficiency is widespread, particularly in tropical Vertisols (Katyal *et al.*, 1987).

Le Mare (1987) has emphasized that the definition of the Vertisol order is based on physical properties, without specifying chemical and morphological criteria. This means that there are no a priori chemical fertility characteristics for Vertisols and these must be evaluated, even for soils already classified as Vertisols.

The usually high nutrient (except for N) status of Vertisols relates to the chemical and mineral composition of the parent materials from which they are developed (Ahmad, 1996). The dominant smectitic clays can be formed by the alteration of base-rich minerals (Dixon, 1982) and their stability is favoured by a high concentration of basic cations and monosilicic acid, as found in basin sites. The fact that smectite-containing Vertisols show shrink–swell properties, according to moisture status, impacts on nutrient requirements and the efficiency of fertilizer use. Even when improved soil water management practices

(eds J.K. Syers, F.W.T. Penning de Vries and P. Nyamudeza)

are implemented successfully, nutrient limitations still remain. However, there are potentially important moisture–nutrient interactions that can be exploited, particularly for N and to a smaller extent for phosphorus (P).

This overview chapter considers specific information on the status and depletion of nutrients in Vertisols, responses to fertilizer nutrient input, and the impact of improved technologies on nutrient requirements. Particular emphasis is directed to Zimbabwe and Ghana, which have two contrasting moisture regimes, with a view to better understanding nutrient–moisture interactions for the development and implementation of sustainable nutrient management practices.

Nutrient Status and Depletion

pH

Vertisols, globally, are mostly neutral to alkaline in reaction. In the Caribbean, Vertisols are acidic, with surface soil pH values in the range of 5.0–6.2 (Ahmad, 1983), whereas in the Sudan values range from 8.0 to 9.5 (Robinson *et al.*, 1970). More commonly, surface soil pH values are 7.0–8.0 (Dudal, 1965). Continuous cultivation of Vertisols can lead to an increase in surface soil pH values because of the bringing up of higher-pH subsoil material (Hussein *et al.*, 1992).

Organic carbon

Because soil organic matter (OM) plays a vitally important role in the retention and cycling of nutrients, the amounts of organic carbon (OC) in Vertisols will be considered briefly. Organic carbon levels are usually low in Vertisols, particularly when they are cultivated continuously. This is the case in spite of their dark colour, which is thought to be due to the formation of organic matter–smectite complexes (Singh, 1954). In temperate regions and in soils cultivated relatively recently, as in Australia and the USA, high values (2–6%) are obtained (Finck and Venkateswarlu, 1982). In the tropics and for soils with a long history of cultivation, OC levels can be very low, e.g. 0.3% in India (Singh, 1954) and in the Gezira soil in the Sudan (Robinson *et al.*, 1970). Intermediate values of about 1.1% were reported by Hussein *et al.* (1992) for cultivated and uncultivated soils in Zimbabwe. In this study there were no major differences in OC between four Vertisols which had been under cultivation for 2, 6, 8 and 20 years, compared with their respective control sites. Elsewhere, there are reports of decreases in OC with continuous cultivation, e.g. in Australia (Cook *et al.*, 1992). OM is usually more or less uniformily distributed with depth in Vertisols.

Nitrogen

As indicated previously, there is a widespread deficiency of N in tropical Vertisols (Katyal *et al.*, 1987). The amounts of N are largely related to the OM content and range from very low surface layer values (0.02–0.06% in the Central Sudan (Ayoub, 1986)) to about 0.08% in Indian Vertisols, with values rarely exceeding 0.1% (Dudal, 1965). Values in excess of 0.1% are reported in Australia and North America (Texas). On the Accra Plains of Ghana, N values are usually lower than 0.08% (Owusu-Bennoah and Duah-Yentumi, 1989). Nyamudeza (1998) determined total N in the surface horizons of Vertisols in Zimbabwe (Fig. 3.1) which had been under continuous cultivation for up to 39 years. The uncultivated soil contained 0.09% N and this level appeared to be maintained in fields cultivated for less than 10 years. For soils cultivated for 30–39 years, there was a substantial decrease in total N to about 0.05%. Whereas OM levels are usually expected to decrease exponentially with cultivation, eventually reaching a steady state reflecting soil type, climate and management practices, there is no evidence for this in the N status of Zimbabwean Vertisols. In Southern Queensland, Australia, total N declined linearly in six soils over 25 years of cultivation and cereal cropping (Dalal and Mayer, 1986a). In the same Vertisols, dry matter yield and N uptake by winter cereal crops also showed significant decreasing trends with period of cultivation (Dalal and Mayer, 1986b).

Smectite minerals have the ability to fix both ammonium (NH_4^+) and potassium (K^+) ions. Although fixed NH_4^+ values of up to 0.3 cmol(+) kg^{-1} have been reported for Gezira soils (Said, 1973), these values were independent of cropping and fertilizer practices and unlikely to have much practical significance for the supply of N (Finck and Venkateswarlu, 1982).

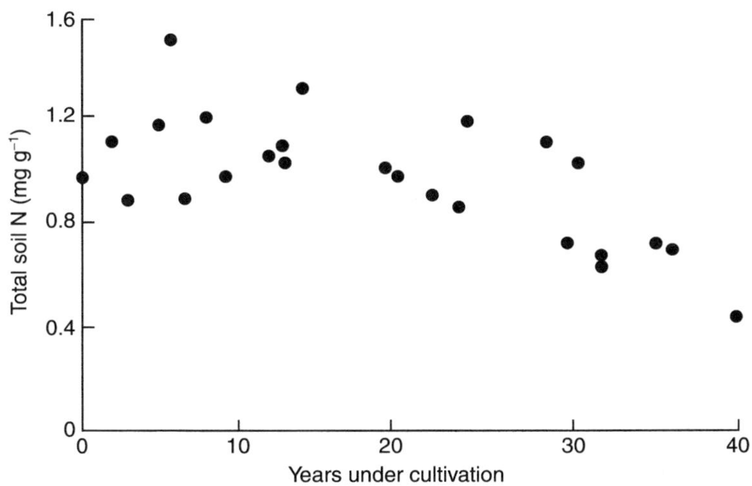

Fig. 3.1. Changes in soil N with years under cultivation in the upper 15 cm of Vertisols in the Chisumbanje area of Zimbabwe. (Data from Nyamudeza, 1998.)

Phosphorus

There is conflicting information in the literature regarding P status and its significance in Vertisols. What is clear is that, next to N, P is usually the most limiting nutrient, but that responses to added P are variable. Results presented by Le Mare (1987) for responses to fertilizer indicate that P is not usually limiting in crop production. In contrast, Swindale (1982) considers that P deficiency is widespread. In Ghana, Vertisols are regarded as being deficient in P, based on soil test results and the soils have high P-sorption capacities (Acquaye and Owusu-Bennoah, 1989). Given these differences it is surprising that there is so little information on the chemistry of P in Vertisols.

The availability of P in Vertisols is usually assessed using the Olsen bicarbonate procedure (Olsen *et al.*, 1954). Values as low as 0.1 mg kg^{-1} have been reported but with a common range of 2–10 mg kg^{-1} (Katyal, 1978). Responses to added P are usually lower in Vertisols of basaltic origin because of their higher P status than those developed on granites and sedimentary parent materials. Calcium-bound P (Ca-P) is the dominant form of P in these near-neutral or calcareous soils and the dissolution of any primary apatite or secondary Ca-P would be very slow under these conditions. In the high OM-containing Vertisols of Texas, organic P is an important source of available P and some soil series show no response to fertilizer P if the C : P of the soil OM in the A horizon is less than 75.

Finck and Venkateswarlu (1982) have pointed to the likely impact of the physical movement of soil on the distribution of available P in Vertisols. Movement of P-enriched soil down cracks could affect the response to added P, both directly and residually. However, results from Zimbabwe show no evidence of downward movement of P, at least from the 0–15 to the 15–30 cm depth (Table 3.1). There is a trend for higher Olsen P values with increasing number of years under cultivation, probably due to uptake of P from depth and return to the soil surface in crop and animal residues.

It is essential to recognize the importance of water in P availability in Vertisols. In flooded and irrigated soils, ferric iron in hydrous ferric oxide,

Table 3.1. Olsen P values (mg kg^{-1}soil) at two depths (0–15 and 15–30 cm) in Vertisols at Chisumbanje Experimental Station (Zimbabwe) as influenced by the number of years under cultivation. (Data from Nyamudeza, 1998.)

		Olsen P at:	
Age of fields (years)	Number of fields	0–15 (cm)	15–30 (cm)
Uncultivated		0.2	0
1–10	7	1.1(±0.14)	0.1(±0.07)
11–20	6	3.2(±0.65)	1.2(±0.62)
21–30	5	4.1(±0.62)	0.4(±0.36)
31–40	7	3.4(±0.69)	0.2(±0.14)

which strongly adsorbs P, is reduced to the ferrous form, decreasing the ability of the soil to adsorb P and increasing P availability (Willett and Muirhead, 1984). However, Turner and Gilliam (1976a) considered that, as most of the inorganic P, at least in alkaline Vertisols, is associated with Ca, the reduction of ferric iron is unlikely to fully account for the observed lower responses of rice in flooded Vertisols than in similar upland soils. Turner and Gilliam (1976b) concluded that the improved availability of P in flooded soils was due to improved diffusivity of P caused by the decrease in soil tortuosity. Anion-exchange resin-extractable P was higher in a saturated soil than in a moist soil, but this effect occurred before reduction of the ferric iron had occurred.

It may be concluded that little is known about the chemistry of P in Vertisols but there is good evidence to suggest that P behaviour differs from that in other soils. Apparently P-deficient soils at ICRISAT, with Olsen-extractable P value down to 1.5 mg kg^{-1}, showed no response with pigeonpea (Burford *et al.*, 1989). Also usually high P-responding sorghum generally only responds to P application on Indian Vertisols when the Olsen P value is less than 2 mg kg^{-1}, which is a very low value when 5 mg kg^{-1} is usually thought of as 'possibly adequate'. The release of small amounts of P through a large volume of soil may provide adequate P for a deep-rooting crop (Le Mare, 1987). Further work is required on P dynamics in Vertisols in contrasting environments.

Potassium

Potassium (K) deficiency is rare in Vertisols. Whereas the total K content of many Vertisols is approximately 1% and the amount usually related to clay content, values for exchangeable K are more variable and commonly range from 1.0 to 13 cmol(+) kg^{-1} (Finck and Venkareswarlu, 1982). Zimbabwean Vertisols contain between 0.2 and 24 cmol(+) kg^{-1} of exchangeable K (Hussein *et al.*, 1992; Nyamudeza, 1998).

Fixation of K has been studied in Indian and Sudanese Vertisols, in particular; fixation is a property related to the content of interstratified smectitic and micaceous clay minerals (Dixon, 1982). Between 2.5 and 25 cmol(+) K kg^{-1}, with a mean of 7.5 cmol(+) K kg^{-1}, was fixed by 22 Vertisols in Rajasthan, India (Bhatnagar *et al.*, 1973), which is a substantial amount of K in relation to the amount of exchangeable K usually present. In the Sudanese Vertisols, where fixed K levels are about 8 cmol(+) kg^{-1}, Finck (1962) showed that K fixation was not sufficiently severe to render K totally unavailable to plants. Said (1971) had previously shown that some of the K released to the exchangeable form on drying was fixed when the soil was re-moistened. The conversion of part of the exchangeable K to 'non-exchangeable K' provided a useful slowly-released source of K from this buffering reservoir and it may explain why K responses are limited, even when exchangeable K levels are low.

Additionally, the high cation-exchange capacity of Vertisols (commonly 30–60 cmol(+) kg^{-1}) also serves to reduce K losses by leaching.

Other

Secondary nutrients (sulphur, calcium and magnesium) have received less attention than N, P and K because they rarely limit crop production. Zinc deficiency can be common in Vertisols (Katyal *et al.*, 1987). Other micronutrient deficiences do occur, for example iron in Tanzania (Le Mare, 1959), but they are not common and tend to be location specific.

Response to Fertilizers

Because deficiencies of nutrients, except for N and P, are rare on Vertisols, responses to these other nutrients will not be considered here.

Nitrogen

Nitrogen is invariably the most limiting nutrient in Vertisols and it is not surprising that most work has been done with this nutrient. Early work by Crowther (1954) in the Sudan Gezira demonstrated the large response of cotton to added N. However, it is dangerous to generalize about fertilizer N responsiveness, given that soil N status varies and responses are very much influenced by moisture status and soil depth (Katyal *et al.*, 1987).

In the semi-arid tropics (SAT), farmers usually do not apply fertilizer N unless the soils are irrigated. For example, in the semi-arid lowveld of Zimbabwe farmers have been cultivating Vertisols continuously for the last 50 years without applying fertilizer N. They argue that due to the low rainfall (500–700 mm) application of chemical fertilizers can increase water stress to their crops. This argument has been supported by research that has shown decreases in dry matter and grain yield of maize with fertilizer use when water is limited (Bennett *et al.*, 1989) and in pearl millet (Affholder, 1995), although in both cases the soils were not Vertisols. On-station research at the Chisumbanje Experimental Station in Zimbabwe in fact showed no significant response of maize to fertilizer N during three seasons, although there was a 17% yield increase in 1989/90 (Table 3.2). In contrast, detailed studies at ICRISAT (India) in 1982, 1983, 1984 and 1985, with seasonal rainfall of 516, 913, 485 and 322 mm, respectively, demonstrated that sorghum yields could be increased by 300–2000 kg grain ha^{-1} on deep Vertisols by the addition of 90 kg N ha^{-1}. Up to 60 kg N ha^{-1}, the response was almost linear (Katyal *et al.*, 1987). These workers considered that, contrary to popular belief, there is 'remarkable potential for increasing crop yields through N fertilization of crops grown on Vertisols over contrasting environments of the SAT'.

The use of improved moisture conservation practices, in the form of tied furrows, resulted in a response of maize to fertilizer N at four sites in 1995/96 and at two of the four sites in 1996/97 in Zimbabwe (Table 3.3). The sites were selected on the basis of the number of years under cultivation, ranging from 5 to 40 years. Maize grain yield responded to fertilizer N in both seasons when rainfall was average to above average. The responses appear to have been influenced by both rainfall and number of years under cultivation.

Sugarcane responded to fertilizer N under irrigation at Chisumbanje Experimental Station, Zimbabwe (Table 3.4). Plant cane gave similar yields for all N rates (60–240 kg ha^{-1}). Nitrogen responses started in the first ratoon (although not significant) and increased as the number of ratoons increased. The lack of response in the first year is an indication of the high nutrient status of the soil. Removal of large quantities of dry matter quickly resulted in a response in the second year.

Under rainfed conditions dry matter removal of sorghum and maize is minimal in most years. Hence under rainfed conditions, crop responses to N

Table 3.2. Response of maize grain (t ha^{-1}) to fertilizer N at Chisumbanje Experimental Station (Zimbabwe). (Unpublished data from Nyamudeza.)

	Added N (kg N ha^{-1})			
	0	50	100	Rainfall (mm)
1985/86	1.93	1.79	1.72	551
1986/87	0.32	0.34	0.31	278
1989/90	5.48	6.24	6.39	532

Table 3.3. Response of maize grain (t ha^{-1}) to fertilizer N, annual rainfall, and years under cultivation (YUC) for different sites in Zimbabwe in 1995/96 and in 1996/97.

			Added N (kg ha^{-1})				
Site	YUC	Rainfall (mm)[a]	0	40	80	120	160
1995/96							
Matikwa	5	640	3.1	4.1	4.0	4.3	4.0
Chisumbanje	10	640	2.6	3.4	4.1	4.4	4.6
Moyana	32	557	1.6	2.8	3.4	3.0	3.5
Machona	39	707	1.3	2.7	3.8	4.7	3.6
1996/97							
Chisumbanje	11	515	4.5	5.3	5.6	5.5	6.1
Machona	40	No record	0.9	1.7	2.7	2.9	3.3

[a]October to April.

Table 3.4. Response of sugarcane (t ha^{-1}) under irrigation to fertilizer N at Chisumbanje Experimental Station, Zimbabwe.

Added N (kg ha^{-1})	Plant cane	First ratoon	Second ratoon	Third ratoon	Mean
60	167.5	115.6	102.7	93.4	119.8
120	168.3	117.8	111.9	109.2	126.8
180	168.4	119.4	108.3	112.4	127.1
240	167.8	119.8	109.0	109.6	126.6
Quadratic effect	NS	NS	$P < 0.01$	$P < 0.001$	

may not occur until after several years, resulting in farmers not appreciating the benefit of fertilizer N.

Phosphorus

Most of the work on crop responses to fertilizer P has been done in India. When yields are low in the SAT, responses to P are usually small but responses can be large and significant, as with rainfed sorghum on granitic Vertisols in Hyderabad and Rajkot, India (Singh and Venkateswarlu, 1985). Under irrigated conditions and particularly with increasing applications of fertilizer N, good responses to P have been obtained in the Sudan (Ayoub, 1986) and in India (Mathau *et al.*, 1978).

Responses to fertilizer P on the Accra Plains in Ghana have been small, even though very low (0.1–3.5 mg kg^{-1}) Olsen-bicarbonate P values are obtained (Acquaye and Owusu-Bennoah, 1989). In this respect, the Ghanaian Vertisols are similar to their Indian counterparts, where responses are often only obtained at Olsen P values lower than 2 mg kg^{-1}, when 5 mg kg^{-1} is the usual value for other Indian soils.

Research in Ghana has shown that the removal of surplus soil surface water by land shaping has a substantial beneficial effect on crop yield (Asiedu *et al.*, Chapter 11). Important interactions between fertilizer rate and land shaping have emerged, with implications for optimum rates of fertilizer application and water quality. A cambered-bed furrow system with 50% of the recommended fertilizer application rate gave a higher yield of maize than 100% of the recommended rate on a flat seedbed. The effect of landform at comparable fertilizer rates shows (Table 3.5) a consistent superiority of the cambered-bed furrow over the Ethiopian-bed furrow, which in turn was superior to the ridge–furrow. The increased grain yield due to landform was larger than that due to fertilizer. Supporting soil and plant analyses showed that P availability and efficiency of use by the crop was the largest in the cambered-bed furrow system, lowest on the flat bed, and intermediate with the Ethiopian and the ridge and furrow systems. Similarly, the use of tied furrows

Table 3.5. Grain yield of maize as influenced by landform at Kpong Agricultural Research Station, Ghana, in 1995.

		Percentage increase in grain yield (%)	
Landform	Fertilizer rate[a]	Effect of landform at comparable fertilizer input	Effect of fertilizer input with comparable landform
Cambered bed	0	73	0
	50	99	128
	100	43	185
Ethiopian bed	0	64	0
	50	48	79
	100	29	258
Ridge	0	20	0
	50	35	122
	100	22	252
Flat	0	0	0
	50	0	98
	100	0	246

[a]Percentage of recommended rate of 250 kg ha^{-1} of 15–15–15 plus 125 kg ha^{-1} of urea.

Table 3.6. Interaction of N and P in the yield of sorghum in the peninsular region of India. (Data from Venkateswarlu, 1980, cited in Finck and Venkateswarlu, 1982.)

Treatment (kg ha^{-1})	Yield (kg ha^{-1})
N_0 P_0	1940
$N_{60}P_0$	2190
N_0P_{60}	2630
$N_{60}P_{60}$	3730

to enhance moisture conservation has given 6-year mean crop yield increases of 26, 27 and 45% for sorghum, cotton and maize, respectively, on Vertisols in Zimbabwe (Nyamudeza *et al.*, Chapter 10), with implications for nutrient offtake and fertilizer requirements. This is seen to be an important area for further work.

Interactions between N and P have been little studied, although it is clearly an important issue. Venkateswarlu (1980, cited in Finck and Venkateswarlu, 1982) demonstrated that, where P is deficient, it is important to correct this, otherwise the response to fertilizer N, per se, may be very small. This is well illustrated by the results from the peninsular region of India (Table 3.6). Further work is required to evaluate N-nutrient interactions.

Needs and Opportunities for Improved Nutrient Management

Farmers in many Vertisol areas have limited options for making easy improvements in nutrient management. Variations in year-to year response to fertilizer nutrients, particularly to N, are determined largely by rainfall. Under rainfed conditions, smallholder farmers do not usually apply fertilizers and in Zimbabwe this extends to both organic and inorganic fertilizers. Where continued cultivation is occurring without any nutrient input, nutrient mining will inevitably contribute to a decline in soil quality and eventually to reduced crop yields. Appropriate nutrient management strategies are required in these situations.

Return of nutrients by the recycling of crop residues is not always feasible because of competing demands, particularly where animal fodder is required. Crop residues left in the field to be eaten by livestock are an important pathway for nutrient return. Incorporation and not burning of crop residues should be encouraged.

Comprehensive studies in India showing the benefits of fertilizer N use (Katyal *et al.*, 1987) have included the improvements in N efficiency that can be obtained with banding and by varying the N source. For example in deep Vertisols, the application of urea using a split-band method gave approximately 55% use efficiency, which was regarded as 'rather high'. On shallow Vertisols the leaching of nitrate-containing fertilizers appeared to be serious.

Improved soil water management technologies – for example improved moisture conservation in semi-arid Zimbabwe (with a mean annual rainfall of as low as 400 mm) and removal of excess surface water, using land-forming techniques on the sub-humid (900–1400 mm) Accra Plains in Ghana – bring about substantial increases in crop yield, with attendant increases in nutrient removal. The technologies also enhance the efficiency of fertilizer nutrient use, with the prospect of obtaining higher yields for a given level of nutrient input, or maintaining crop yield with a reduced nutrient input. This is seen to be an important area of further research and there is a need for impact assessment analysis of increased yields through the adoption of improved surface soil management technologies.

Response to fertilizer N in a wide range of environments is variable over time and there is a need to establish improved maintenance application rates and frequencies that accommodate, as far as possible, the risk element already highlighted.

Uncertainties remain regarding the Olsen test on Vertisols. Low values of about 2 mg P kg^{-1}, without a concomitant response to added P, are puzzling and the need for further studies on P behaviour and dynamics remains as clear now as it was in 1987 (Le Mare, 1987).

Under irrigation, the demand for N, P and K is larger and if there are moves to increase the area of irrigated Vertisols, as in Zimbabwe, it will be necessary to improve fertilizer recommendations. The priority will inevitably be with N.

Conclusions

Vertisols are inherently fertile soils, having a near-neutral to alkaline pH (6.5–8.0) and high nutrient status, except for N, which is universally deficient in tropical Vertisols.

Despite their dark colour, Vertisols usually contain low amounts of OC (0.3–1.0%), particularly if they are cultivated continuously. In comparison, N levels are usually in the range 0.02–0.08% N and rarely exceed 0.1%. The smectitic clay minerals in Vertisols can fix NH_4^+ (and K^+), but it seems unlikely that fixed NH_4^+ has much practical significance for N supply.

Phosphorus is the next most limiting nutrient after N in Vertisols but there is conflicting information on the extent of P deficiency. Moisture status is important in influencing P availability, being higher in flooded soils. Our understanding of P behaviour in Vertisols is poor and further work is required, on P dynamics in particular.

Potassium deficiency is rare and slowly-exchangeable K, resulting from K fixation, may serve as an important buffer for K in Vertisols.

Responses to fertilizer N vary and are influenced by moisture status. Some researchers consider that there is a remarkable potential for fertilizer N use in the SAT, while farmers in several countries remain to be convinced, pointing to the need for more effective developmental research and demonstration work.

Responses to P are very variable, even though Olsen P values are usually very low, as in much of India. Interactions with moisture are again important and in sub-humid areas (as in Ghana) there are very good opportunities for increasing crop yield by using land-shaping practices that increase the availability and efficiency of P use.

Farmers in many Vertisol areas have limited options for substantially improving their management of nutrients. The effective recycling of nutrients in crop residues is hindered by competing uses. Incorporation of any crop residues not consumed by animals must be promoted. In the absence of mineral fertilizer input, continued nutrient mining is inevitable. Nutrient depletion, resulting from continuous removal of crop products, poor recycling of residues and often limited input of chemical fertilizers, is a major concern, particularly in sub-Saharan Africa. Varying the form of fertilizer N according to soil (particularly depth) and rainfall conditions offers some potential.

There are continuing uncertainties regarding the significance of the Olsen P values with Vertisols but it is with N–moisture interactions where the best prospects for improvement of crop yields and N use efficiency lie. Inefficient use of fertilizer N occurs with a deficit and a surplus of water. There is a need to optimize N and water use efficiency and in this context modelling offers very good potential.

References

Acquaye, D.K. and Owusu-Bennoah, E. (1989) Chemical characteristics of some Vertisols of Ghana. In: *Vertisols Management in Africa*. IBSRAM Proceedings No. 9, IBSRAM, Bangkok, Thailand, pp. 328–346.

Affholder, F. (1995) Effect of organic matter input on the water balance and yield of millet under tropical dryland conditions. *Field Crops Research* 19, 297–311.

Ahmad, N. (1983) Vertisols. In: Wilding, L.P., Smeck, N.E. and Hall, G.F. (eds) *Pedogenesis and Soil Taxonomy*. II. Soil Orders. Elsevier Science, Amsterdam, pp. 91–123.

Ahmad, N. (1996) Management of Vertisols in rainfed conditions. In: *Vertisols and Technologies for their Management*. Developments in Soil Science 24, Elsevier, Amsterdam, pp. 363–428.

Ayoub, A.T. (1986) ^{15}N-labelled urea recovery by different crops in the Sudan Gezira soil. *Fertilizer Research* 9, 213–221.

Bennett, J.M., Mutti, L.S.M., Rao, P.S.C. and Jones, J.W. (1989) Interactive effects of nitrogen water stresses on biomass accumulation, nitrogen uptake and seed yield of maize. *Field Crops Research* 19, 297–311.

Bhatnagar, R.K., Natham, G.P., Chouhan, S. and Seth, S.P. (1973) Potassium fractions and fixing capacity of medium black soils of Rajasthan. *Journal of the Indian Society of Soil Science* 21, 429–432.

Burford, J.R., Sahrawat, K.L. and Singh, R.P. (1989) Nutrient management in Vertisols in the Indian semi-arid tropics. In: *Management of Vertisols for Improved Agricultural Production. Proceedings of an IBSRAM Inaugural Workshop*. International Crops Research Institute for the Semi-Arid Tropics, Patancheru, AP, India, pp. 147–157.

Cook, G.D., So, H.B. and Dalal, R.C. (1992) Structural degradation of two Vertisols under continuous cultivation. *Soil and Tillage Research* 24, 47–64.

Crowther, F. (1954) A review of experimental work. In: *Agriculture in Sudan*. Oxford University Press, Oxford.

Dalal, R.C. and Mayer, R.J. (1986a) Long-term trends in fertility of soils under continuous cultivation and cereal cropping in Southern Queensland. V. Rate of loss of N from the soil profile and changes in C : N ratios. *Australian Journal of Soil Research* 24, 493–504.

Dalal, R.C. and Mayer, R.J. (1986b) Long-term trends in fertility of soils under continuous cultivation and cereal cropping in Southern Queensland. I. Overall changes in soil properties and trends in winter cereal yields. *Australian Journal of Soil Research* 24, 265–279.

Dixon, J.B. (1982) Mineralogy of Vertisols. In: *Vertisols and Rice Soils in the Tropics. Transactions of the 12th International Congress of Soil Science*, International Soil Science Society, New Delhi, India, pp. 48–60.

Dudal, R. (1965) Dark clay soils of tropical and subtropical regions. *Agricultural Development Paper* No. 83, FAO, Rome.

Finck, A. (1962) Kalium status of some Sudan clays. *Plant and Soil* 16, 293–309.

Finck, A. and Venkateswarlu, J. (1982) Chemical properties and fertility management of Vertisols. In: *Symposium Papers II.Transactions of the 12th International Congress of Soil Science*, International Soil Science Society, New Delhi, India, pp. 61–79.

Hussein, J., Adey, M.A. and Elwell, H.A. (1992) Irrigation and dryland cultivation effects on the surface properties and erodibility of a Zimbabwe Vertisol. *Soil Use and Management* 8, 97–103.

Katyal, J.C. (1978) Management of phosphorus in lowland rice. *Phosphorus in Agriculture* 73, 21–34.

Katyal, J.C., Hong, C.W. and Vlek, P.L.G. (1987) Fertilizer management in Vertisols. In: *Management of Vertisols under Semi-Arid Conditions.* IBSRAM Proceedings No. 6, IBSRAM, Bangkok, Thailand, pp. 247–266.

Le Mare, P.H. (1959) Soil fertility studies in three areas of Tanganyika. *Empire Jounal of Experimental Agriculture* 27, 197–222.

Le Mare, P.H. (1987) Chemical fertility characteristics of Vertisols. In: *Management of Vertisols under Semi-Arid Conditions.* IBSRAM Proceedings No. 6, IBSRAM, Bangkok, Thailand, pp. 125–137.

Mathau, K.K., Sankaran, K., Nanakabushui, N. and Krishnamoorthy, N. (1978) Effect of continuous rotational cropping on the organic carbon and total nitrogen content in a black soil. *Journal of the Indian Society of Soil Science* 26, 283–285.

Nyamudeza, P. (1998) Water and fertility management for crop production in semi-arid Zimbabwe. PhD thesis, The University of Nottingham, UK.

Olsen, S.R., Cole, C.V., Watanabe, F.S. and Dean, L.A. (1954) Estimation of available phosphorus in soils by extraction with sodium bicarbonate. *US Department of Agriculture Circular* No. 939. USDA, Washington, DC.

Owusu-Bennoah, E. and Duah-Yentumi, S. (1989) A report of a survey of current land use and cropping systems of the Vertisols of the Accra Plains. In: *Vertisols Management in Africa.* IBSRAM Proceedings No. 9, IBSRAM, Bangkok, Thailand, pp. 399–406.

Robinson, G.H., Magar, W.Y. and Rai, K.D. (1970) Soil properties in relation to cotton growth. In: Seddig, M.A. and Hughes, L.C. (eds) *Cotton Growing in Sudan Environment.* Agricultural Research Council, Republic of Sudan, Wad Medani, 318 pp.

Said, M.B. (1971) Effect of drying and rewetting on the exchangeable potassium content of the Sudan Gezira soil. *Agrochimica* 15, 356–361.

Said, M.B. (1973) Ammonium fixation in the Sudan Gezira soils. *Plant and Soil* 38, 9–16.

Singh, R.P. and Venkateswarlu, J. (1985) Role of all-India coordinated research project for dryland agriculture in research development. *Fertilizer News* 30, 67–80.

Singh, S. (1954) A. study of black cotton soils with special reference to their coloration. *Journal of Soil Science* 5, 289–299.

Swindale, L.D. (1982) Distribution and use of arable soils in semi-arid tropics. In: *Managing Soil Resources. Transactions of the 12th International Congress of Soil Science,* International Soil Science Society, New Delhi, India, pp. 67–100.

Turner, F.T. and Gilliam, J.W. (1976a) Effect of moisture and oxidation status of alkaline rice soils on the absorption of soil phosphorus by an anion resin. *Plant and Soil* 45, 353–363.

Turner, F.T. and Gilliam, J.W. (1976b) Increased P diffusion as an explanation of increased P availability in flooded rice soils. *Plant and Soil* 45, 365–377.

Willett, I.R. and Muirhead, W.A. (1984) Nutrient transformations in a grey clay soil during rice rotation in relation to fertilizer requirements of post-rice crops. In: McGarity, J.W., Hoult, E.H. and So, H.B. (eds) *Reviews in Rural Science* No. 5. University of New England, Armidale, Australia, pp. 121–135.

New Tools for Research and Development to Promote Sustainable Land Management

4

E.T. CRASWELL[1] AND
F.W.T. PENNING DE VRIES[2]

[1]*International Board for Soil Research and Management (IBSRAM);*
[2]*IWMI (formerly of IBSRAM), Bangkok, Thailand*

Key words: research and development, sustainability, soil management, technology transfer, information science, models

Introduction

More effort is needed globally to help solve the problems of poor land users, who must cope with uncertain land tenure, inadequate institutions for decision making, and negative economic on- and off-site impacts of land degradation. Particularly vulnerable are those farmers cultivating marginal lands. These farmers have been the neglected clients of international agricultural research. For them, soil fertility decline is a major problem because they cannot afford the inputs required to make agriculture sustainable. Land degradation needs to be stopped or reversed in those marginal areas that still have sufficient potential and in those areas where degradation itself has major off-site impacts (erosion, water pollution). There is also a need for monitoring of trends in degradation, and for devising early warning systems where thresholds are approached (Penning de Vries, 2000). The most challenging goal for the research and development community is to achieve sustainable land management (SLM), particularly in the hot spots or most vulnerable areas (Scherr and Yadav, 1995).

Many successes have been recorded in research to develop improved technologies to overcome land degradation and improve productivity. Research on drainage and water conservation through improved surface management of Vertisols, the focus of the Harare workshop on which this book is based, provides some examples. Waterlogging is a major constraint to crop production in Vertisol areas of Ethiopia. The broadbed maker developed and tested through collaboration between the International Livestock Centre for Africa, Institute of Agricultural Research (IAR), International Crops Research Institute for the Semi-Arid Tropics and International Board for Soil Research and Management (IBSRAM) (Abebe and Jutzi, 1989) increased crop yields significantly and has now been adopted by thousands of farmers in that country (Schioler, 1998). On the other hand Nyamudeza *et al.* (1992) showed that, in the dry areas of Zimbabwe, tied ridges improve water conservation and hence the yields of maize, sorghum and cotton crops. Some local farmers were convinced by these findings and adopted the technology, although the availability of traction and other factors significantly affect the rate of adoption, particularly among poor farmers. Thus technologies for the management of Vertisols have had some success, but many farmers in relevant Vertisol areas in Africa and on other continents are being bypassed by the fruits of research.

Our ultimate objective must be to induce not just hundreds and thousands but many millions of farmers to practice SLM. Rates of adoption of the knowledge-intensive technologies required for land management are generally lower than adoption rates of new high-yielding varieties and agrochemicals, which have been characterized by Röling (1990) as hardware. Traditional soil science has failed to tailor scientific knowledge to the needs of clients such as extension workers and farmers. There is an urgent need to address the knowledge gap by collecting, organizing and making available relevant SLM knowledge for better decision making on land management in developing countries. Furthermore, we need to examine the effectiveness and impact of research on land management and overcome shortcomings by devising and testing new approaches and tools.

This chapter examines the problems and failures of past research on land management and new paradigms for research developed to overcome those problems. We then review emerging new tools for prioritizing and targeting research, measuring sustainability, assessing research impact, and scaling up to promote adoption. Finally the chapter discusses the consortium model as a framework for organizing the activities of research institutions around a common goal.

The Poor Track Record of Land Management Research

Human-induced land degradation is a serious global problem, particularly in marginal areas where low rainfall, declining soil fertility, acidity, steep slopes, salinity and poor soil structure are major constraints to agriculture (Oldeman *et al.*, 1990; Scherr and Yadav, 1995). Degradation and declining productivity continue unabated despite the fact that knowledge of soils and their management abounds in the scientific literature. In contrast, indigenous knowledge about sustainable land management has not been well documented; the publication by Pongsapich (1998) is one of a number of exceptions. The lack of impact of research on land management is a major dilemma facing practitioners and advocates of a scientific approach to combating land degradation. Craswell and Niamskul (1999) listed some of the past problems that have contributed to the lack of success of research. Firstly, researchers have ignored the needs of clients. Furthermore, there is clearly a breakdown in the line of communications between land users and policy makers on the one hand, and the scientific community on the other. Often farmers are blamed for land degradation and for not adopting technologies promulgated by researchers but not relevant to the needs of farmers. A top-down approach to extension lies behind low rates of adoption. The non-governmental organizations (NGOs) have championed a grass roots approach, but their work is generally isolated from the mainstream extension and research organizations supported by governments. At the other end of the spectrum, policy makers have not been well informed by researchers about the need to improve the policy and legislative environment in which farmers cultivate marginal lands. Turning to the research itself, one of the main problems has been the reductionist training of many soil scientists. This has led to too much emphasis on small plots and on strategic research into processes in isolation from the target farming systems. Furthermore, biophysical scientists have been slow to develop collaboration with colleagues in the fields of economics, sociology and anthropology that hold the key to ensuring relevant and client-oriented research.

Framework for Using the New Research Tools

The lack of success in promoting SLM, particularly in marginal areas, points to the need for an overhaul in approaches to research on SLM. A new research paradigm and other tools are described below in detail. However, firstly the chapter outlines a framework for the use of the tools in the context of the goal of SLM (see Fig. 4.1).

The starting point for effective SLM research is to define the problems of unsustainability and set priorities for a focused effort. This can be done effectively using the resource management domain (RMD) concept. As shown in Fig. 4.1, the extent and nature of hot spots or high-priority target areas are defined first. This may be done on the basis of a serious land degradation problem, e.g. soil erosion on steep lands or acid, infertile soil. An alternative would be the unproductive use of Vertisols due to waterlogging or drought. In each case social and economic dynamics would be key dimensions in the definition of the target RMD, to ensure that technological solutions defined were transferable to other areas within the same RMD. Some widespread problems of an important RMD may occur in several countries within a region or be scattered across the globe (see Fig. 4.2 for the scale dimensions). In the case of Vertisol management problems, opportunities to improve soil surface management to minimize waterlogging occur within a biophysically defined

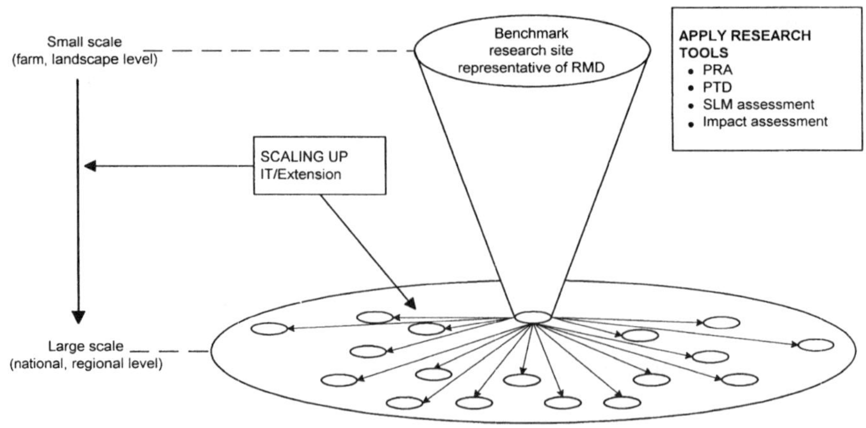

Fig. 4.1. Scaling up from small scale and detailed participatory research to broad scale adoption using the resource management domain (RMD) concept and new research tools from the IBSRAM global tool kit. PRA, participatory rural appraisal (includes gender analysis); PTD, participatory technology development; SLM, sustainable land management.

RMD that includes land in many countries in Africa, South Asia, Oceania and the Americas. However, when social and economic criteria such as small farm size, low labour costs and availability and limited capitalization are included in the identification of the RMD, the geographic scope of the RMD shrinks to the Vertisols in Africa and South Asia.

As shown in Fig. 4.1, detailed research is then conducted in an area(s) representative of the RMD. These areas may be communities or watersheds, depending on the nature of the problem. Key tools used in the detailed research include participatory rural appraisal and participatory technology development, which are an integral part of the research paradigm described below. Technologies and practices based on both scientific and indigenous knowledge are tested for sustainability using the Framework for Evaluation of Sustainable Land Management (FESLM). This tool can also be used for problem definition. Furthermore, the on- and off-site impacts of the land management practices must be assessed *ex ante* or *ex post* at the representative RMD to ensure, before scaling up the technology, that it meets the sustainability criteria.

Technologies and SLM knowledge can then be scaled up to reach the thousands/millions of smallholders farming within the target RMD. Scaling up requires information dissemination through demonstration, investment and extension. The complexity of knowledge-intensive technologies for many land management systems lends itself to dissemination through modern information technology systems (see Fig. 4.1). Decision support systems (DSS) may be necessary to transfer knowledge to land managers and policy makers.

All of the steps in detailed research on representative RMDs and the scaling up to national or regional scales require a clear understanding of

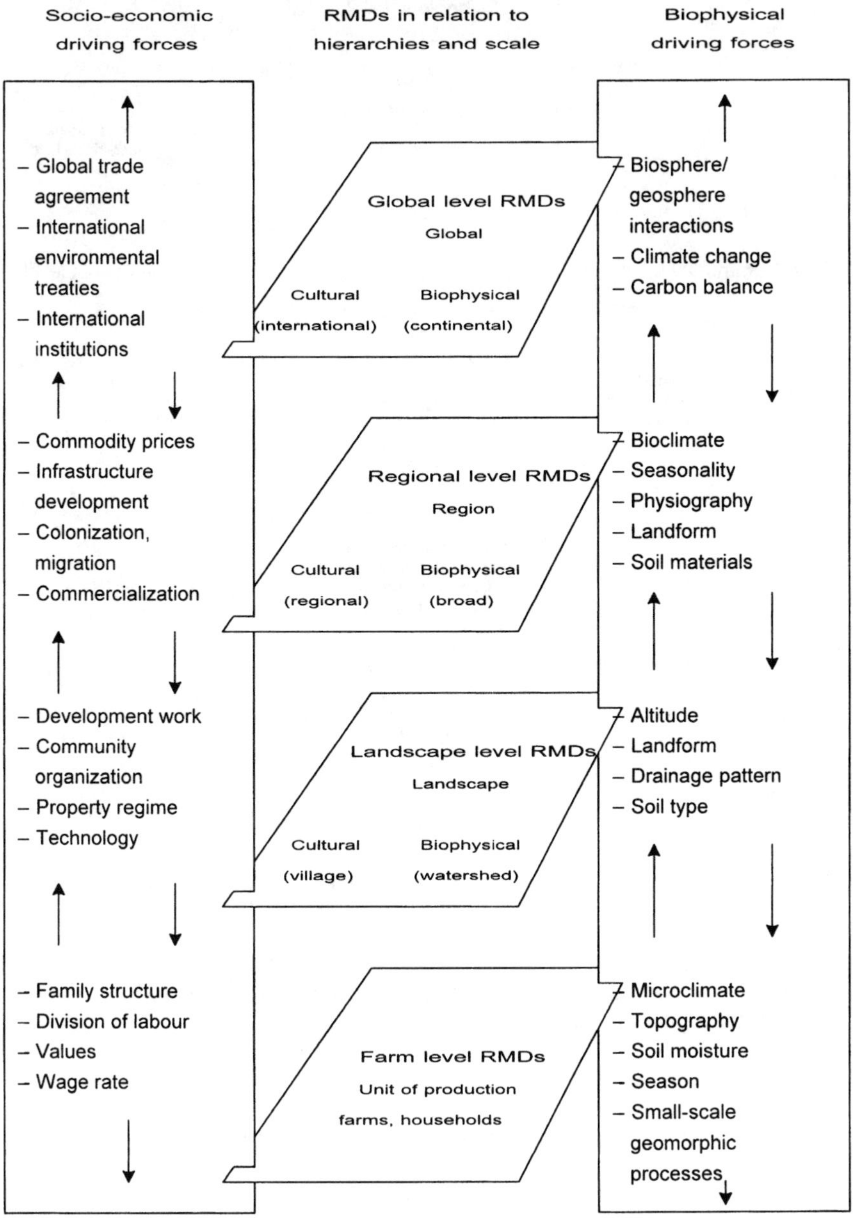

Fig. 4.2. Relationships among scale, land-use change and resource management domains (RMDs) (Dumanski and Craswell, 1998).

the interactions between the multiple components of SLM: i.e. technology, institutions, social networking and policy/legal levels (Fig. 4.3). Effective SLM technology development and scaling up depend on successful interventions at

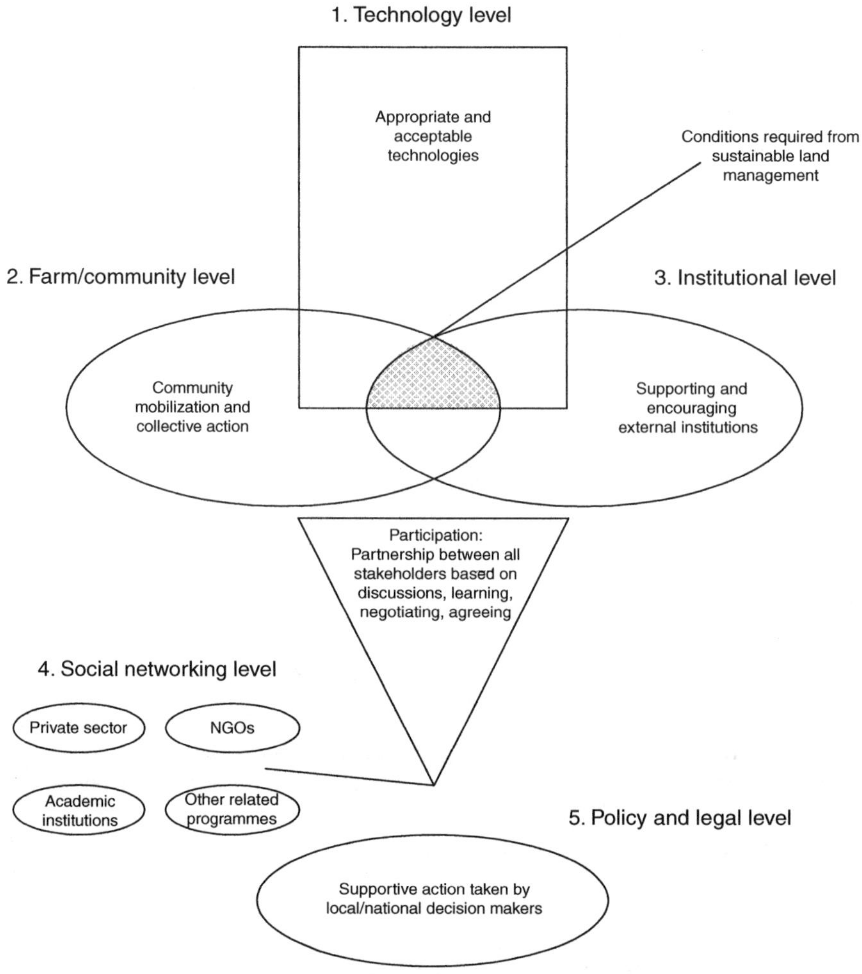

Fig. 4.3. Basic components for sustainable land management. NGOs, non-governmental organizations.

all levels. The frameworks presented in Fig. 4.1, 4.2 and 4.3 provide a means for addressing SLM problems successfully. Detailed descriptions of some of the research tools in Fig. 4.1 are now presented.

A New Research Paradigm

Greenland *et al.* (1994) undertook a major review of the past successes and failures of research on soil, water and nutrient management, and devised a new agenda based on innovative concepts about how to improve its effectiveness and impact. The new research paradigm they expounded is based on an interdisciplinary approach and has the following characteristics:

- *User orientation* – participatory and community based at all stages from planning to implementation.
- *Policy* – focused on policy and institutional issues that influence farmer and community decisions.
- *Equity* – concerned with equity, including gender analysis, in research planning and implementation.
- *Landscape* – integration of people, land and water at every scale from plot to catchment.
- *Research intensity* – linking strategic, applied and adaptive research with technology development and participatory dissemination.
- *Knowledge* – reliance on both indigenous and scientific information.
- *Goals* – linking increased productivity with natural resources conservation.

Resource Management Domains

Effective planning of research on natural resources requires a framework for characterizing farming systems and their impact on the natural resource base. The framework can be used to set priorities and to transfer knowledge and technologies between similar areas. Craswell *et al.* (1999) point out that, historically, biophysically based units based on natural potential have dominated land management and land-use planning. The RMD breaks with this tradition by recognizing the key role of human decisions and activities and environmental consequences. The concept of RMDs implies human interventions in management of natural resources, particularly rural landscapes. Traditionally, the potential for this has been analysed using biophysical variables, but increasingly it is being realized that socio-economic information has an equally important role (Dumanski and Craswell, 1998).

> **Definition of an RMD:** A resource management domain (RMD) is a spatial (landscape) unit that offers opportunities for identification and application of resource management options to address specific issues. It is derived from georeferenced biophysical and socio-economic information, and it is dynamic and multiscale in that it reflects human interventions in the landscape.

The potential uses include technology transfer, technology generation, prioritization of research/research planning, land-use planning and policy programme formulation and implementation. Potential users include national and international agencies, agribusinesses, farmer associations, etc.

The potential users of RMD methodology/concepts could be from local, regional or national agencies to international and multi-national agencies such as United Nations agencies and international research organizations. The tools needed in studying RMDs would range from simple thematic mapping tools to sophisticated modelling, simulation, geographical information systems (GIS) and knowledge-based systems (Rais *et al.*, 1997). These information technology tools will help in identifying RMDs and relating the information to problem domains such as SLM.

The vehicle for delivering RMD output/information to users would vary according to the needs of the users: it may be a simple printed leaflet or a sophisticated on-line GIS-based RMD information system (RMD-IS). The RMD-IS would require trained users as well as computer scientists to manage it. At its simplest aggregate level, the RMD for Vertisols in sub-Saharan Africa at the farm and landscape level could be described as follows:

- heavy clay soils with vertic properties;
- tropical, semi-arid to sub-humid climate;
- low infiltration, erosion-prone;
- prone to waterlogging;
- subsistence farmers;
- women responsible for many farming operations;
- low yields;
- low external inputs and declining soil fertility;
- low cost and availability of labour;
- uncertain land tenure; and
- poor infrastructure and distant from markets.

The kind of information and data needed by users to use an RMD-IS system varies with the intended scale. In addition, the complexity of the RMD in terms of the physiographic conditions, the land-use systems, the performance levels, the values of the community, and interactions with other RMDs and sectors of society further complicate the information needs and the ability to characterize the RMDs.

Modern tools are available for processing the kinds and quantities of information available for understanding the systems and evaluating the responses. Each of the tools has limitations and specific needs that must be appreciated to be effectively used. The tools available include:

- GIS, models and expert systems, relational databases;
- optimization techniques;
- hierarchical decision making;
- techniques for data aggregation, disaggregation;
- expert systems technology; and
- decision support systems.

Assessing Sustainability

Sustainability and sustainable development are important concepts debated at length in the wake of the 1987 release of the Brundtland Commission Report –

Our Common Future (World Commission on Environment and Development, 1987). The central tenet is that development should *meet the needs of the present without compromising the ability of future generations to meet their own needs.* Becker (1997) surveyed a broad sweep of ethical, cultural, economic and environmental issues that impinge on sustainability assessment, concluding with a plea for honesty, modesty, clarity and caution. The FESLM, developed by an international interdisciplinary team led by the Food and Agriculture Organization and IBSRAM (Smyth and Dumanski, 1993), embodies many of these principles, while providing a practical basis on which sustainability can be assessed in the field. The FESLM sets the goal of a system that combines technologies, policies and activities aimed at integrating socio-economic principles with environmental concerns so as to simultaneously: maintain or enhance production/service (productivity); reduce the level of production risk (security); protect the potential of the resource base (protection); be economically viable (viability); and be socially acceptable (acceptability). The objectives constitute the five pillars of the FESLM.

Field testing of the FESLM began in 1995 in steep-land areas of Cebu, in the Philippines (Gomez *et al.*, 1996). The field team found that the choice, measurement and interpretation of sustainability indicators presented considerable practical problems. They rearranged the FESLM to focus on indicators of farmer acceptance and on resource conservation. Soil indicators such as soil depth, soil organic carbon and groundcover proved particularly useful in characterizing differences between farms of adopters and non-adopters of soil conservation practices. Radar graphs provide a means of effectively displaying the many dimensions of the FESLM (Fig. 4.4). In this case, the farm that had adopted contour hedgerow systems on the steep land exceeded the trigger or threshold values for all of the indicators except crop failure, for which a farm registering a lower value was more sustainable. The Cebu work underlined the importance of farmer participation in the process of sustainability assessment and the need for an interdisciplinary approach. Both are central elements of the new paradigm for research on soil, water and nutrient management discussed above.

Assessing Impact

Impacts of land management on the resource base and environment are expressed on site and often off site. Assessing these impacts and the role of research and technology in enhancing positive impacts, or minimizing negative impacts, presents difficult challenges. Nevertheless, information about impacts is essential as a basic research management tool and as a basis for persuading donors and decision makers to invest in research on the problems. Recognizing this need, IBSRAM has recently started a programme to develop tools for assessing the impacts of soil erosion and nutrient decline. Enters (2000) developed a framework for assessing the off-site impacts of soil

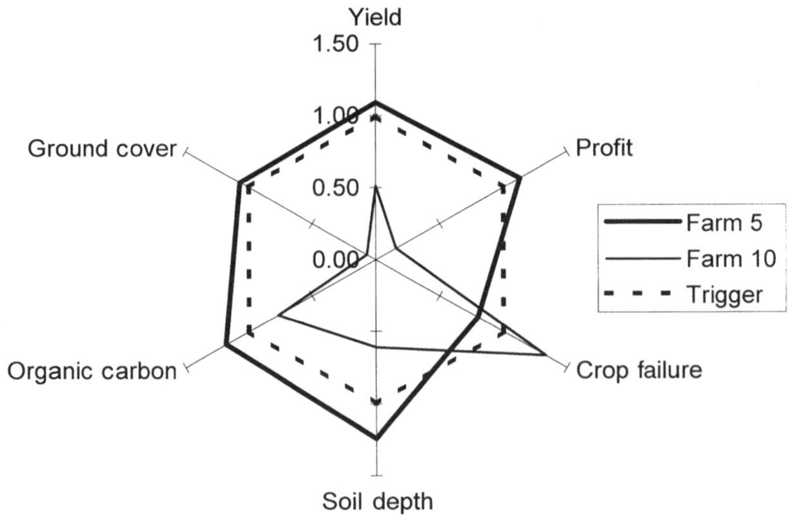

Fig. 4.4. The sustainability cobweb for key Framework for Evaluation of Sustainable Land Management parameters for two farms in Cebu – Farm 5 adopted soil conservation measures and Farm 10 did not. The trigger is the threshold for sustainability.

erosion. Enters' paper illustrates the complexity of the task and the paucity of studies of off-site impacts. On-site damage due to soil loss is difficult to separate from yield decline due to concurrent nutrient depletion. Off-site impacts include the off-site benefits to low-lying farms on which silt is deposited, and increased runoff volume that accrues to reservoirs. Off-site damage includes flash-flooding and siltation of reservoirs. The siltation impacts of intensified cultivation in upper catchments are difficult to separate from natural erosion. Significant reductions in estimates of the life of water reservoirs are common in developing countries because increases in siltation due to soil erosion in the catchment combine with the common practice of overestimating the life of water storages. Cost–benefit analysis is the preferred method for valuing the costs of soil erosion, but valuation methodologies are far more advanced than our understanding of the environmental impacts.

The impact of nutrient depletion also has on- and off-site dimensions. Drechsel and Gyiele (1999) reviewed the analytical issue to develop a framework for the economic assessment of nutrient depletion. Their use of the nutrient balance model with a replacement cost approach indicates that the costs of nutrient depletion in some sub-Saharan African countries are as high as 20% of the agricultural gross domestic product. Drechsel and Gyiele go further to develop a framework for multi-criteria analysis that integrates results from replacement cost analysis, productivity change approach, total factor productivity and farmer assessment. This more comprehensive approach provides a broader umbrella that, in contrast to cost–benefit analysis, accounts for non-monetary costs and benefits such as sustainability. The next step in this work is to develop a framework for impact analysis of both soil erosion and nutrient depletion.

Information Technology

Fortunately recent developments in the field of information science provide tools such as DSS and GIS for coping with these complexities. The World Overview of Conservation Approaches and Technologies Consortium provides a large database of tested technologies, particularly from Africa. Another example is the prototype DSS based on the FESLM that IBSRAM recently developed with Agriculture Canada. The prototype is undergoing alpha testing by extension workers and researchers in steep-land areas of Asia. The so-called DSS-SLM distils knowledge from special case studies and IBSRAM network experiments into a rule base consisting of 25 questions or indicators covering all five pillars of the FESLM. A computer program assesses the data entered and interprets the degree of sustainability of the subject land management system. Planned refinements include a prescriptive module that will suggest remedies in the case of unsustainable land management, and a GIS dimension that will help define RMDs at different scales from the field to the catchment level. Both the FESLM and the DSS require further rigorous testing and development, but each has the potential to enhance significantly our capacity to assess sustainability.

To answer better the needs for information and knowledge on SLM in developing countries, one or more knowledge centres for sustainable land management should be established, to make available and accessible data sets, methods, tools, research and training capacity on SLM and assessment of sustainability. A land management compendium could provide a vehicle for making such knowledge widely available.

Institutional Roles

National agricultural research and extension systems (NARES) should be further reinforced in their capacity to undertake effective natural resource management research, to make use of land management expertise from other countries (South–South exchange) and from advanced research institutes (ARIs) (North–South exchange), and to cater to the needs of different classes of stakeholders (including women, minorities). NGOs increase in strength, so that it is practical to listen to their voice in R&D planning to identify 'issues', and to involve them as active partners in research. The growth of the private sector as part of the problem and as part of the solution in unsustainable land management requires active exploration of how to collaborate with the private sector.

IAR Centres play an important role.

> Participants in global programmes benefit in many ways including: (1) global prioritization of research needs; (2) improved possibilities for funding for programme participants through the recognition of the programme by donor agencies; (3) close interaction with, and knowledge of, other research teams within their area of specialization; (4) opportunities for interdependent research projects (i.e. projects requiring interdisciplinary and complementary partnerships); (5) improved access to information and resources; and (6) participation in programme meetings and conferences.
>
> (Frison *et al.*, 1997)

> The private sector plays a major global role in agricultural R&D. . . . In the future, private sector investments in agriculture and food are expected to increase faster than investments by the public sector, in both industrial and developing countries. . . . noted that, as farmers use more purchased inputs and as the value-added in agriculture increasingly moves off the farm to the marketing and processing subsectors, it is likely that the incentives for private sector investments in agricultural research will grow. With current private sector global revenues in fertilizers, seeds, pesticides, and animal health alone estimated conservatively at approximately $70 billion per year, the private sector is an essential partner for the global public sector engaged in agricultural research.
>
> (James, 1996)

Natural resources research is largely regarded in the area of national and international public goods. Poor countries can expect little of this development in the near future, and there is a risk of accelerated degradation by profit taking in big companies in the food and industrial sector; there is also a lack of organized public 'green concerns' in many countries. Nevertheless, some agrochemical companies provide nutrients and herbicides that can contribute significantly to SLM. Their research rarely targets smallholders in marginal areas.

NARES need to be supported to ensure that all their information is fully used when it comes to land management, and that they contribute their knowledge and data. The gap between info-have and info-have not should become smaller. Technical developments make it possible to make all data and knowledge accessible through a single window, independent of where the data physically are. This allows for the use of distributed databases. A centre on SLM information, therefore, can provide single-window access to information from all over the world, and be a virtual centre.

The Consortium Model

The organizational model that is most suited to the new research paradigm, and that accommodates the complexity of interdisciplinary work, is the consortium which draws together research agencies around a common goal or high-priority SLM theme. At the core of the consortium are the farmers, NARES and NGOs. These core participants are supported by ARIs and IAR

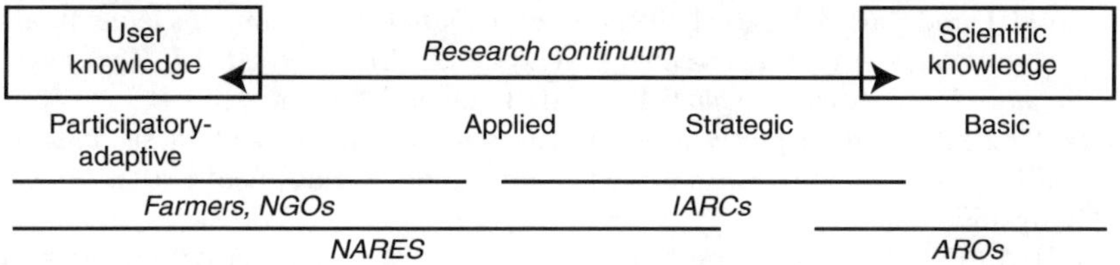

Fig. 4.5. The organization of a consortium showing primary stakeholder domains along the research continuum. NGOs, non-governmental organizations; IARCs, International Agricultural Research Centres; NARES, national agricultural research and extension systems; AROs, advanced research organizations.

Centres, each contributing according to their comparative advantage (see Fig. 4.5).

Conclusions

The innovative tools described in this chapter signal a significant departure from traditional soil science research. The tools have not been tested widely and much hard work lies ahead to apply them to the intractable problems of land degradation and to make fine adjustments to them. These steps must be taken to improve the livelihoods of the millions of rural poor farming marginal lands and to arrest the decline in the land resource base. The Vertisols of Africa present an excellent opportunity for such work.

References

Abebe, Mesfen and Jutzi, S. (1989) The Joint Project on improved management and utilization of dark clay soils in Ethiopia – retrospects and prospects. In: Ahn, P.M. (ed.) *Vertisol Management in Africa.* IBSRAM Proceedings No. 9, IBSRAM, Bangkok, Thailand, pp. 385–398.

Becker, B. (1997) Sustainability assessment: a review of values, concepts, and methodological approaches. *Issues in Agriculture* No. 10.

Craswell, E.T. and Niamskul, C. (1999) Watershed management for erosion control on sloping lands in Asia. In: Lal, R. (ed.) *Integrated Watershed Management in the Global Ecosystem.* CRC Press, Boca Raton, Florida, pp. 65–72.

Craswell, E.T., Rais, M. and Dumanski, J. (1999) Resource management domains as a vehicle for sustainable development. *International Journal on Failures and Lessons Learned in Information Technology.* Special Issue on Information System Development (Vol. 2, No. 1).

Drechsel, P. and Gyiele, L. (1999) *The Economic Assessment of Soil Nutrient Depletion: Analytical Issues for Framework Development.* Issues in Sustainable Land Management No. 7, IBSRAM, Bangkok, Thailand.

Dumanski, J. and Craswell, E.T. (1998) Resource management domains for evaluation and management of agro-ecological systems. In: *International Workshop on Resource Management Domains*. IBSRAM, Bangkok, Thailand, pp. 1–13.

Enters, T. (2000) *Methods for the Economic Assessment of the On- and Off-Site Impacts of Soil Erosion*, 2nd edn. Issues in Sustainable Land Management No. 2, IBSRAM, Bangkok, Thailand.

Frison, E.A., Collings, W.W. and Sharrock, S.L. (1997) Global programmes: a new vision in agricultural research. *Issues in Agriculture* No. 12.

Gomez, A.A., Swete-Kelly, D.E., Syers, J.K. and Coughlan, K.J. (1996) Measuring sustainability of agricultural system at the farm level. In: *Methods of Assessing Soil Quality*. Special Publication No. 9, Soil Science Society of America, Madison, pp. 402–407.

Greenland, D., Bowen, G., Eswaran, H., Rhoades, R. and Valentin, C. (1994) *Soil, Water, and Nutrient Management Research – a New Agenda*. IBSRAM Position Paper, IBSRAM, Bangkok, Thailand.

James, C. (1996) Agricultural development and research: the need for public–private seater partnerships. *Issues in Agriculture* No. 9.

Nyamudeza, N., Mazhangara, E. and Kamba, E. (1992) Adoption of the tied furrow technique, and the effects of the technique and the previous crop on the residual moisture and yields of sorghum and maize on Vertisols. In: *Report of the 1992 Annual Review Meeting on AFRICALAND Management of Vertisols in Africa*. Network Document No. 3, IBSRAM, Bangkok, Thailand, pp. 69–82.

Oldeman, L.R., Hakkeling, R.T.A. and Sombroek, W.G. (1990) *World Map Status of Human Induced Soil Degradation. Global Assessment of Soil Degradation (GLASOD)*. ISRIC, Wageningen, The Netherlands.

Penning de Vries, F.W.T. (2000) Land degradation: information needs and challenges to research. In: Bridges, E.M. *et al.* (eds) *Response to Land Degradation*. Oxford University Press, New Delhi, India.

Pongsapich, A. (1998) *Indigenous Technical Knowledge for Land Management in Asia*. Issues in Sustainable Land Management No. 3, IBSRAM, Bangkok, Thailand.

Rais, M., Craswell, E.T., Gameda, S. and Dumanski, J. (1997) Decision support system for evaluating sustainable land management in sloping lands of Asia. In: *Conference on Geo-information for SLM*, 17–21 August 1997, Enschede, The Netherlands.

Röling, N. (1990) The agricultural research-technology interface: a knowledge systems perspective. In: Kaimowitz, D. (ed.) *Making the Link: Agricultural Research and Technology Transfer in Developing Countries*. Westview Press, with International Service for National Agricultural Research, Boulder, Colorado, pp. 1–42.

Scherr, S.J. and Yadav, S. (1995) *Land Degradation in the Developing World: Implications for Food, Agriculture and the Environment to 2020*. Food, Agriculture and the Environment Discussion Paper No. 14, IFPRI, Washington, DC, 36 pp.

Schioler, E. (1998) *Good News from Africa – Farmers, Agricultural Research and Food in the Pantry*. International Food Policy Research Institute, Washington, DC.

Smyth, A.J. and Dumanski, J. (1993) *FESLM: an International Framework for Evaluating Sustainable Land Management*. World Soil Resources Report No. 73, FAO, Rome.

Part II

Country Papers and National Perspectives on the Management of Vertisols

Vertisols Management in Malawi

5

A.D.C. CHILIMBA

Chitedze Agricultural Research Station, Lilongwe, Malawi

Key words: East Africa, Malawi, Vertisols, soil management, soil and water conservation, tillage, soil degradation

Introduction

Background information

Malawi is a long narrow landlocked country situated between latitudes 9°22′ and 17°8′ south of the equator and longitudes 33°40′ and 35°55′ east. Its total area is 11.8 million ha, of which 9.4 million ha is land and 2.4 million ha is covered by water; 2.2% of the total land area is covered by Vertisols (Table 5.1). Malawi is divided into eight Agricultural Development Divisions as shown in Fig. 5.1. Blantyre Agricultural Development Division has the highest area under Vertisols (10.76%). Salima, Shire Valley, Karonga and Machinga Agricultural Development Divisions have 4.75%, 2.60%, 0.65% and 0.34% of their areas covered with Vertisols, respectively (FAO, 1998). The Vertisols are found in low-lying areas where the mean minimum temperature is 13.4°C and the mean maximum temperature is 37.5°C; they experience low average annual rainfall of 711–813 mm.

Agriculture is the mainstay of Malawi's economy, accounting for more than 35% of the gross domestic product; it employs 80% of the total labour force, and contributes about 90% of the domestic export earnings. Approximately 85% of the total population reside in rural areas and derive their livelihoods from farming (NSO, 1987). Smallholder farmers traditionally practised shifting cultivation, with soil fertility being rejuvenated by long fallow periods. However, pressure from rapidly increasing human population has led to reduced fallow periods and continuous cultivation with little or no added external inputs. Under such circumstances, smallholder agriculture causes a serious increase in soil erosion, surface runoff, nutrient depletion, overgrazing, deforestation, diminishing groundwater supplies and loss of biological

Table 5.1. Soils of Malawi and their estimated areas based on the soil map of Malawi. (Source: Lowole, 1995.)

Soil name (FAO)	Land area (km^2)	% of total land area
Ferric Luvisols	6,636.3	7.0
Lithosols	23,138.7	24.5
Ferric Luvisols with Lithosols	6,551.7	6.9
Lithosols with some Luvisols	1,207.3	1.3
Orthic Ferralsols	3,175.0	3.4
Orthic Ferralsols, Chromic Luvisols	4,990.7	5.3
Xanthic Ferralsols	14,849.0	15.5
Orthic Ferralsols and Xanthic	3,868.0	4.1
Ferralsols	3,729.6	4.0
Xanthic Ferralsols over massive laterite	4,214.0	4.5
Xanthic Ferralsols and Lithosols	1,222.7	1.3
Humic Ferralsols	861.3	0.9
Humic Ferralsols with Lithosols	615.2	0.7
Dystric Nitosols	699.8	0.7
Dystric Nitosols with some Lithosols	1,399.5	1.5
Dystric Nitosols and Lithosols	1,945.5	2.1
Pellic Vertisols	144.4	0.1
Chromic Vertisols	201.0	0.2
Calcic Phaeozems	1,222.7	1.3
Orthic Solonetz	307.6	0.3
Eutric Regosols	8,797.2	9.3
Eutric Fluvisols, Eutric Cambisols	3,875.7	4.1
Eutric Gleysols	599.8	0.6
Eutric Fluvisols and Eutric Gleysols	94,253.2	99.6
Totals		

diversity. Low and/or declining soil fertility is the single most important factor responsible for reduced crop yields in the smallholder sector (Saka *et al.*, 1995).

Soil degradation

Soil degradation is responsible for the loss of soil productivity through chemical, physical and biological processes. Physical degradation involves destruction of soil structure, increase in soil bulk density, a decrease in soil water-holding capacity of the soil and a decrease of infiltration rates. These physical constraints lead to increased surface runoff, which causes soil erosion. Vertisols are undoubtedly among the most fertile soils found in Malawi (Table 5.2). Soil degradation in Malawi has resulted in food shortages, reduced exports and increased food imports, which drain foreign exchange reserves. This has very serious implications for food security and the country's economy. Malawi is facing a potentially serious soil degradation problem arising from current inappropriate land use, continuous cultivation, deforestation and

Fig. 5.1. Agricultural Development Divisions (ADDs).

cultivation of marginal lands, such as steep slopes, shallow soils and stream banks. National average soil erosion is estimated at 20 t ha^{-1} $year^{-1}$, accounting for mean yield losses of between 4 and 11% of maize $year^{-1}$, and is the most serious threat to sustained crop production (Anon., 1998). Soil loss due to soil erosion is deposited in the streams/rivers and lakeshore plains, causing serious siltation of water bodies in the country. Some eroded soils are deposited on lower lands and are used for growing crops.

Table 5.2. Soil analytical data for two soil profiles of Vertisols (Mphonde) in Shire Valley Agricultural Development Division. (Source: Panje, 1979.)

Depth (cm)	pH ($CaCl_2$)	Silt (%)	Clay (%)	Texture	% C	S (mg kg^{-1})	P (mg kg^{-1})	K (cmol(+) kg^{-1})	Na (cmol(+) kg^{-1})	Ca (cmol(+) kg^{-1})	Mg (cmol(+) kg^{-1})
0–15	7.5	10	58	Clay	1.49	0	8	0.60	0.36	43.8	8.11
15–30	7.5	10	57	Clay	0.98	11	5	0.70	0.52	59.4	8.32
30–45	7.6	10	60	Clay	0.95	6	3	0.47	0.56	46.9	9.15
0–15	7.5	9	53	Clay	1.64	2	3	0.60	0.33	48.8	7.07
15–30	7.5	8	58	Clay	1.01	27	3	0.60	0.38	56.3	7.45
30–45	7.5	12	51	Clay	0.85	2	3	0.68	0.42	51.3	7.90

Vertisol characteristics

Vertisols in Malawi are derived from recent colluvial lacustrine and fluvial deposits. They are very deep, imperfectly to poorly drained, dark grey soils and have sandy clay to clay texture. The soil shrinks and becomes hard during the dry season and swells and becomes sticky and plastic during the rainy season, thus making cultivation very difficult. They are severely ponded in the rainy season and may experience flash floods during periods of high rainfall. Problems in managing Vertisols are caused by the aggregates of clay being quite unstable in water. The soil colloids shatter immediately when wetted and the resulting fine particles fill the cracks with topsoil. Since the soil slakes so easily when it rains, it causes close packing and rainfall infiltration is retarded, resulting in excessive accumulation of surface runoff. The movement of the excessive surface runoff causes soil erosion and sometimes results in floods (Jones, 1973). However, the cracks increase rainwater infiltration into the subsoil. The accumulated subsoil water makes the clay swell, generating sub-soil pressures, which lead to deformation of the surface soil (known as gilgai relief). The soil structure is strong, coarse sub-angular blocky to prismatic. The soil reaction is slightly alkaline to alkaline (pH 7.0–8.5), total nitrogen is medium (0.12–0.25%), available phosphorus is low to medium (< 8 mg kg^{-1}), exchangeable potassium is medium to high (0.8–1.0 cmol(+) kg^{-1}), and the cation exchange capacity is medium to high (10–20 cmol(+) kg^{-1}). The soil has high exchangeable sodium (20–50%) and slightly saline 2–4 mS cm^{-1} in some places. Soil chemical data for two profiles of the same soil are given in Table 5.2. The results show that Vertisols are among the most fertile soils in the country.

Soil and water conservation

Water is the most important natural resource and has the greatest impact on agricultural production. Inadequate and poorly distributed rainfall and poor soil fertility are the major constraints to crop production under rainfed agriculture (Nyakatawa, 1996). The World Bank (1992) recommendations for drought mitigation and sustainable strategies were to encourage water infiltration and retention practices that maximize rainfall use in crop production, practices that reduce moisture evaporation from the soil and measures that build up soil organic matter. In Malawi, the organic matter content of a soil dropped significantly in the first year of cultivation and, nationally, crop yields were estimated to be dropping by 2% annually (Elwell and Rook, 1996). Data available in the region show that the current tillage practices contribute significantly to loss of rainfall as surface runoff.

Rainfed agriculture provides an important contribution to the subsistence of smallholders, who represent 85% of farmers in Malawi. These farmers are dependent upon rainfall reliability. In a country with a high human population with high land pressure, the benefits of water conservation could be considerable in increasing crop yields. Malawi suffers from persistent droughts, which greatly affect the yields of many crops produced by smallholders. Boxed or tied ridges increase water infiltration into soil and reduce soil erosion, thereby increasing the profile water content (Hulugalle, 1987; Sanders, 1988; Ashworth, 1990). High levels of crop yields are seldom reached on Vertisols due to various limitations such as tillage difficulties, low infiltration rates and permeability and nutrient deficiencies. The amount and distribution of rainfall in areas where Vertisols occur are extremely variable from season to season. In some years the amount of rain is barely adequate to grow a good crop of cotton, in others, a mid-season drought occurs that reduces production. The solutions for managing Vertisols suggested in Malawi were to encourage buildup of organic matter by the introduction of crop rotations that include deep-rooting legumes and grasses. Where the erosion hazard is high, soil physical conservation backed by agronomic measures is recommended.

Materials and Methods

Land preparation methods were investigated to study the effect of land preparation on the soil water conserved and cotton yields. The main plot size was 14.0 × 10.96 m; the design was a randomized block with six replicates. The main plots had the following treatments: flat cultivation (representing the traditional practice); basins (a rectangular flat plot surrounded on all sides by ridges); and ridges with cross ties. The ridges were 91 cm apart. The plots were split as sub-plots with or without nitrogen application. The seed rate was 8–10 seeds per planting station, at 60 cm apart. Three weeks after emergence, the crop was thinned to three plants per station. Nitrogen was applied at 50 kg N ha^{-1} at first flowering in treatments with nitrogen. The experiment was conducted at Tomali in 1978.

A subsoiling and deep ploughing experiment was conducted at Ngabu to compare the effect of subsoiling, mechanically ploughed and ridged, hand ridged, and flat planting on cotton yields with or without farmyard manure (FYM). The main plot and sub-plot treatments were as shown in Table 5.3.

The FYM contained 2.14% N, 0.33% P and 6.03% K. The FYM was spread on the plots and fertilizer was applied in bands between the rows on flat plots or along the ridge for ridged plots. Three tines, 60 cm long and mounted behind a tractor, were used for subsoiling at the beginning of the season in November.

Monitoring of changes in yields of cotton under continuous cultivation was conducted at Zunde to monitor the changes in crop yields under monocropping compared with plots under Rhodes grass. There have been suggestions that declining yields of seed cotton on Vertisols in Shire Valley were caused by monocropping, among other reasons. The treatments were as follows:

Table 5.3. Subsoiling and deep ploughing experiment. (Source: Panje, 1979.)

Main plot treatments	Sub-plot treatments
Flat planting	Nil treatment farmyard manure (FYM_0)
Subsoil, tractor ploughed and ridged	FYM at 7.53 t ha^{-1}
Tractor ploughed and ridged	Fertilizer applied at 162 kg N ha^{-1} and 25 kg P ha^{-1}
Subsoil and hand ridged	
Hand ridged with cross-ties	

1. Under permanent Rhodes grass, opened for cotton in the fifth year.
2. Under permanent cotton, annually applied with cattle *khola* bedding.
3. Ex-Rhodes grass opened for cotton in the third year.
4. Ex-Rhodes grass opened for cotton in the fourth year.
5. Ex-Rhodes grass opened for cotton in the fourth year and FYM applied at 30 t ha^{-1} in the fourth year and 134 t ha^{-1} in the fifth year.
6. Ex-Rhodes grass opened for cotton in the fourth year but covered with grass mulch.

An investigation of tillage systems for Vertisols was conducted using cotton as a test crop. The treatments consisted of planting on the flat, planting on old ridges and planting on new ridges. The experiment was conducted at Ngabu and Zunde and laid out as a randomized block design with six replications. The plot size was 27.30 m^2 and the ridges were 91 cm apart. Five cotton seeds were planted per station and the stations were 60 cm apart. The seedlings were thinned to three, 3 weeks after emergence.

Results and Discussions

In the experiment on land preparation methods, impeded seedling emergence and water infiltration occurred early in the season. Surface capping and excess runoff were observed. These effects were severe on the flat; stunted growth with few bolls were observed on natural slope plots, where there were abnormal symptoms of retarded seedling establishment (Panje, 1979).

The moisture retention characteristics of Vertisols are given in Table 5.4 and gravimetric water content indicated that water was collected at the lower ends on flat plots (the natural slope) and basin plots (Table 5.5). Seed cotton yields on ridges with cross ties and basins were significantly superior to flat planting on the natural slope in all years (Table 5.6 and Fig. 5.2). However, the yield differences due to nitrogen application were not statistically significant (Table 5.6). Seed cotton yield was significantly increased with the introduction of ridges with cross ties and basins. The current practice of planting on the flat gave the least cotton yields of all the treatments (Fig. 5.2).

Table 5.4. Moisture retention characteristics. (Source: Panje, 1979.)

Depth (cm)	Clay (%)	Silt (%)	Textural class	1/3 bar (%)	15 bar (%)	AWC (%)
0–20	26	20	Scl	22.0	10.6	11.4
20–40	33	19	Scl	22.0	11.7	10.3
40–60	34	16	Scl	18.0	9.75	8.3
60–80	17	11	Sl	12.6	6.45	6.2
80–100	22	5	Scl	12.1	5.8	6.3

AWC, available water content; Scl, sandy clay loam; Sl, sandy loam.

Table 5.5. Effects of land preparation methods on gravimetric water content (% v/w). (Source: Panje, 1979.)

Land preparation	Depth of sampling	Upper position of the plot	Lower position of the plot
Ridges with cross ties	0–15	21.8	16.9
	15–30	10.8	11.5
	30–45	7.3	0
	45–60	7.7	8.7
Basins	0–15	19.5	18.6
	15–30	14.4	11.2
	30–45	10.1	10.7
	45–60	10.7	10.1
Flat (natural slope)	0–15	10.1	13.7
	15–30	10.8	12.4
	30–45	9.9	8.8
	45–60	7.8	9.4

Table 5.6. Effect of land preparation and nitrogen application on cotton yields (kg ha^{-1}). (Source: Panje, 1979.)

Land preparation	50 kg N ha^{-1}	0 kg N ha^{-1}	Mean
Ridges with cross ties	1860	1770	1815
Basins	1834	1911	1873
Flat	1117	1032	1075
Mean	1604	1571	

SE = ± 59.5; CV% (main plots) = 29.9; CV% (sub-plot) = 14.7.

The results in the subsoiling and deep ploughing study showed that running water damaged ridged plots, causing considerable soil erosion. Excess water stood for prolonged periods in black gilgai depressions, especially on the flat plots. The yields of subsoil and hand-ridged plots were significantly higher

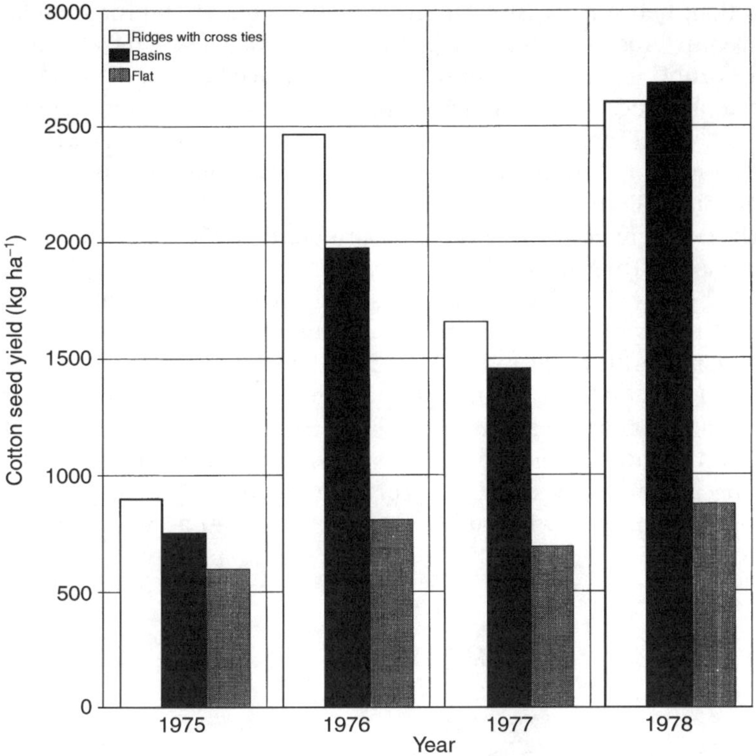

Fig. 5.2. The effect of land preparation methods on cotton seed yield (kg ha^{-1}) across seasons.

Table 5.7. Effects of land preparation and farmyard manure application on seed cotton yields (kg ha^{-1}). (Source: Panje, 1979.)

Method of land preparation	FYM	Zero treatment	Inorganic fertilizer	Mean
Flat planting	1355	1617	1328	1433
Subsoil, tractor plough and ridged	1720	1046	1252	1339
Tractor plough and ridged	1400	985	1393	1259
Subsoil and hand ridged	1728	1290	1670	1563
Hand ridged with cross ties	1126	1164	1537	1276
Mean	1466	1220	1436	1374

SE = 76.0; CV% (main plots) = 10.3; CV% (sub-plots) = 12.4.
FYM, farmyard manure.

than subsoil, tractor plough and ridged, tractor plough and ridged, and hand ridged with cross ties (Table 5.7). The results show that frequent use of a tractor in land preparation increased soil compaction, which reduced rainfall infiltration, hence increasing surface runoff, which erodes the topsoil. Hand ridged with cross ties and flat planting without subsoiling also increased surface runoff, and rainfall infiltration was retarded because of hardpans. This resulted in reduced cotton yields. FYM and fertilizer application significantly increased yield (Table 5.7). The current continuous cultivation with no inputs might result in nutrient depletion, thereby threatening sustainable crop production on Vertisols.

The experiment on the effect of continuous cultivation on the yield of seed cotton showed that cotton plants under grass for 4 years plus grass mulching remained yellow for a long period; this could be due to immobilization of soil nutrients by micro-organisms as the grass has a higher C : N ratio. The yield of cotton significantly increased with cattle bedding application followed by FYM addition. Fields under grass and opened in the fifth year were the most inferior of all treatments in seed cotton yield (Fig. 5.3). The results show that FYM or cattle bedding application significantly increased cotton yields and that the application of cattle or FYM is crucial in managing Vertisols for increased crop production. Fallowing under grasses resulted in accumulation of grass

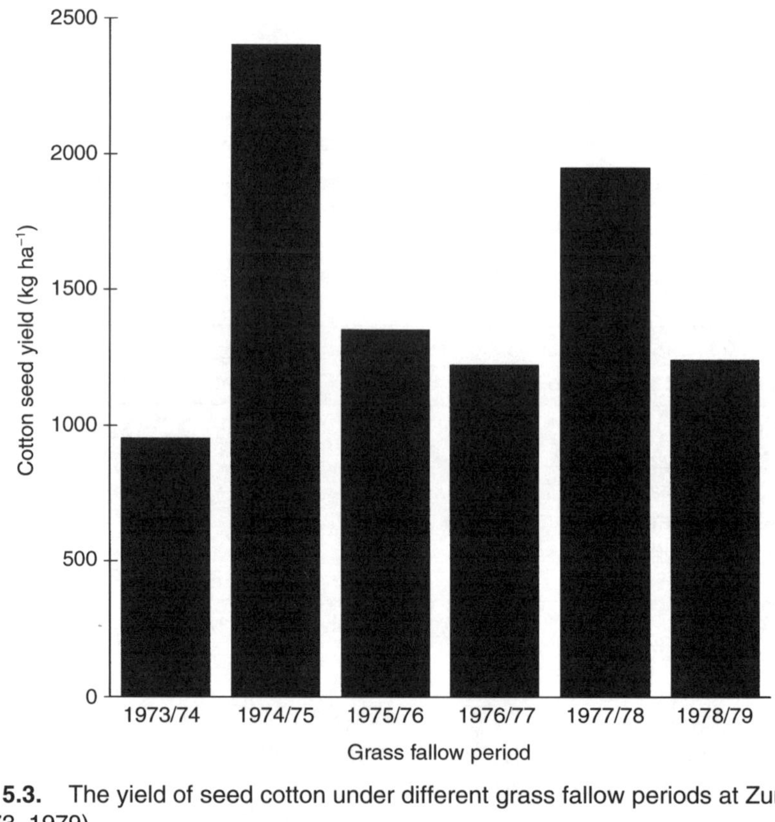

Fig. 5.3. The yield of seed cotton under different grass fallow periods at Zunde (1973–1979).

biomass, which led to nutrient immobilization. Nutrient immobilization might have resulted in reduced yields of cotton.

Results of the tillage systems investigation for Vertisols show that tillage systems affected rooting depth and the infiltration rate at Ngabu. The rooting depth and infiltration rate were lower on the flat than on ridges. However, cotton yields were slightly higher on old ridges than planting on flat or new ridges (Table 5.8). At Zunde, old ridges and new ridges had higher porosity, infiltration rates and higher available water content than planting on the flat. Cotton yields on old ridges and new ridges were slightly higher than planting on the flat (Table 5.9). The infiltration rates were higher at Zunde than at Ngabu because Zunde has more typical Vertisols than Ngabu (with more cracks and an increased infiltration rate). Considering the higher labour requirement for new ridges, permanent ridges are the best options for farmers who are constrained by labour requirements.

Conclusion

The research conducted in Malawi on Vertisols showed that water conservation such as ridges with cross ties, the use of basins, and subsoiling with hand ridging increased cotton yields. Application of cattle manure or FYM also improved cotton yields. Research conducted on Vertisols in Malawi has shown that improved management of Vertisols could lead to increased productivity. Most of Malawi soils are degraded and the soils that remain fertile are the

Table 5.8. The effect of tillage on rooting depth, infiltration rate and cotton yields at Ngabu. (Source: Chilimba, 1989.)

Tillage systems	Infiltration rate ($cm\ h^{-1}$)	Rooting depth (cm)	Cotton yields ($kg\ ha^{-1}$)
Planting on flat	3.27	40	2742
Planting on old ridges	8.38	54	2808
Planting on new ridges	9.41	51	2642

Table 5.9. The effect of tillage on infiltration rates, available water content, porosity, and cotton yield. (Source: Chilimba, 1989.)

Tillage systems	Infiltration rate ($cm\ h^{-1}$)	AWC (%)	Porosity (%)	Cotton yield ($kg\ ha^{-1}$)
Planting on flat	15.63	8.97	49.95	1650
Planting on old ridges	18.87	11.50	56.60	1728
Planting on new ridges	18.94	13.33	56.5	1723

AWC, available water content.

Vertisols. Therefore, there is a need to intensify research on management and nutrient deficiencies in crop production on Vertisols.

References

Anon. (1998) Land resources conservation department activities and achievements 1997/98. A paper presented at the Ministry of Agriculture and Irrigation Planning Workshop, held at Natural Resources College, Lilongwe, Malawi, 23 January, 1998.

Ashworth, V.A. (1990) Agricultural technology and communal farm sector. In: *Zimbabwe Agricultural Sector Memorandum*. Ministry of Agriculture, Harare, Zimbabwe, pp. 1–158.

Chilimba, A.D.C. (1989) Tillage systems for vertisols. In: Chilimba, A.D.C. (ed.) *Annual Report for Soil Fertility and Plant Nutrition, Bvumbwe Research Station*, Limbe, Malawi, 16 pp.

Elwell, H.A. and Rook, J.M. (1996) *Soil and Water Conservation Technology in Four Selected Countries in Southern Africa (Malawi, Mozambique, Zambia and Zimbabwe)*. Vol. 1, Main Report. Agriculture and Environment Division Survey Unit, Lilongwe, Malawi.

FAO (1998) *Soil Fertility Initiative for Malawi*. Report No. 98/036CP-MLW. Lilongwe, Malawi.

Hulugalle, N.R. (1987) Effect of tied ridges on soil water content, evapo-transpiration, and root growth and yield of cowpeas in the Sudan Savanna of Burkina Faso. *Field Crops Research* 17, 219–243.

Jones, E. (1973) Soil Productivity Unit. *Agricultural Research Council of Malawi Annual Report*, Makoka, Zomba, Malawi.

Lowole, M.W. (1995) Soils of Malawi and their inherent constraints. *File D12, E2, Soil Survey Unit*, Lilongwe, Malawi.

NSO (National Statistical Office) (1987) *Malawi Government Monthly Statistical Bulletin*. Government Printer, Zomba.

Nyakatawa, E.Z. (1996) Rainwater and soil fertility management for sustainable cropping on sandy soils in Semi-Arid South East Lowveld of Zimbabwe. In: *Cropping in Natural Region V*. Proceedings of a workshop held in Harare. Agritex and DR&SS, Harare.

Panje, P. (1979) Soil productivity research in Shire Valley. *Bvumbwe Agricultural Research Station Annual Report for the Year 1978 to 1979*. Limbe, Malawi.

Saka, A.R., Green, R.I. and Ng'ong'ola, D.H. (1995) *Proposed Soil Management Action Plan for Malawi*. ODA/WD/MAI, Lilongwe, Malawi.

Sanders, J.H. (1988) Agricultural research and cereal technology introduction in Burkina Faso and Niger. *Agricultural Systems* 30, 139–154.

World Bank (1992) *Malawi Economic Report on Environmental Policy*. World Bank, Lilongwe.

Vertisols Management in South Africa

6

A.J. VAN DER MERWE,[1] M.C. DE VILLIERS,[1] C. BÜHMANN,[1] D.J. BEUKES[1] AND M.C. WALTERS[2]

[1]ARC-Institute for Soil, Climate and Water, Pretoria, South Africa;
[2]ARC-Plant Protection Research Institute, Pretoria, South Africa

Key words: South Africa, Vertisols, soil management, constraints, tillage

Introduction

In South Africa, Vertisols occur throughout the semi-arid and sub-humid summer rainfall zones, occupying just under 2% of the total land area and representing 1.1% of its arable land. Rainfall is erratic over most of South Africa. Low soil water availability is the most serious constraint to crop production on Vertisols and crops experience drought stress at a soil water content as high as 20–25%. Variable crop response, for which Vertisols are conspicuous, is ultimately attributable to low soil water availability.

Vertisols are highly productive, display characteristic physicochemical–biological properties and are of great economic agricultural importance. The Springbok Flats, where 5% of the country's arable Vertisols occur, account for as much as 17% of the total cotton, 7% of the sunflower, 6% of the grain sorghum and 3% of the wheat production.

South Africa's Vertisols are generally underused for food production. Considering the country's annual population growth rate of 2.5% (van der Merwe, 1995) and the limited extent of arable land of high agricultural potential, Vertisols should be brought into optimal production to achieve sustainable rural livelihoods. Underuse, particularly by the rural poor, is ascribed to the difficulty in managing these soils and to crop failure.

In the Springbok Flats over 90% of the area is under arable cultivation. Tillage and cropping markedly affect the structure, morphology and chemistry of Vertisols. Soil degradation, manifested primarily in reduced dry aggregate stability, resulted in increased crusting and in gilgai mounds of increased size and prominence. The general perception of gilgai is that of mounds, depressions and shelves that are observed on tropical and subtropical vertic soils. Gilgai denotes the seasonal impondment of water in shallow depressions and there are structural differences between mound and shelf soils (Grubb, 1993). Large gilgai mounds with a predominantly weak aggregated structure cause poor soil water retention and stunted growth. Runoff from Vertisols in the Springbok Flats is estimated at 20 to 76% and erosion losses at 10 to 12 t ha^{-1} $year^{-1}$. Up to 80% of the precipitation during high-intensity storms can be lost through runoff. Cultivation over some 50 years accounts for a 55% decrease in organic matter content and related decline in yield and biological activity as well as an increase in bulk density (Grubb, 1993).

The objective of this chapter is to review and discuss physical, chemical and biological constraints associated with Vertisol cultivation. Unsustainable cultivation practices induce detrimental changes in soil structure, which could be prevented and even reversed by following sustainable cultivation practices such as residue management, conservation or no-till and crop rotation. Although these practices and their economic benefits have been advocated for almost two decades, the adoption rate by farmers is poor. On-farm demonstration trials and farmer research should facilitate a higher adoption rate.

Distribution and Attributes of South African Vertisols

Vertisols occupy an estimated 2.36 million ha, representing just under 2% of South Africa's surface area (Fig. 6.1). Vertisols correlate particularly with the intermediate weathering products of basic igneous rocks and argillaceous sediments in positions of subdued relief (Bühmann and Schoeman, 1995; Jacobs and Beukes, 1997). Vertisols occur throughout the semi-arid and subhumid summer rainfall zones of South Africa under rainfall regimes that range from 392 to 1353 mm $year^{-1}$, but are particularly common in areas with a mean annual precipitation of 500–700 mm. It is debatable, however, whether Vertisol formation in South Africa reflects climatic conditions or whether it should be ascribed to the outcrop pattern of mafic igneous rock that are concentrated in the 500 to 700 mm rainfall range (Bühmann and Schoeman, 1995). Over most of South Africa, rainfall is erratic and its efficiency on Vertisols is low because of extensive runoff and evaporation during summer months.

The Springbok Flats, selected as the major study area for this chapter, are situated between 28°03′ and 29°16′ longitude east and 24°22′ and 25°08′ latitude south. They constitute part of the volcanic Karoo Supergroup. The area is fairly well sheltered from cold winds. Summer temperatures are high while winters are mild. Rainfall is variable and highly unpredictable and occurs mainly in summer as high-energy showers.

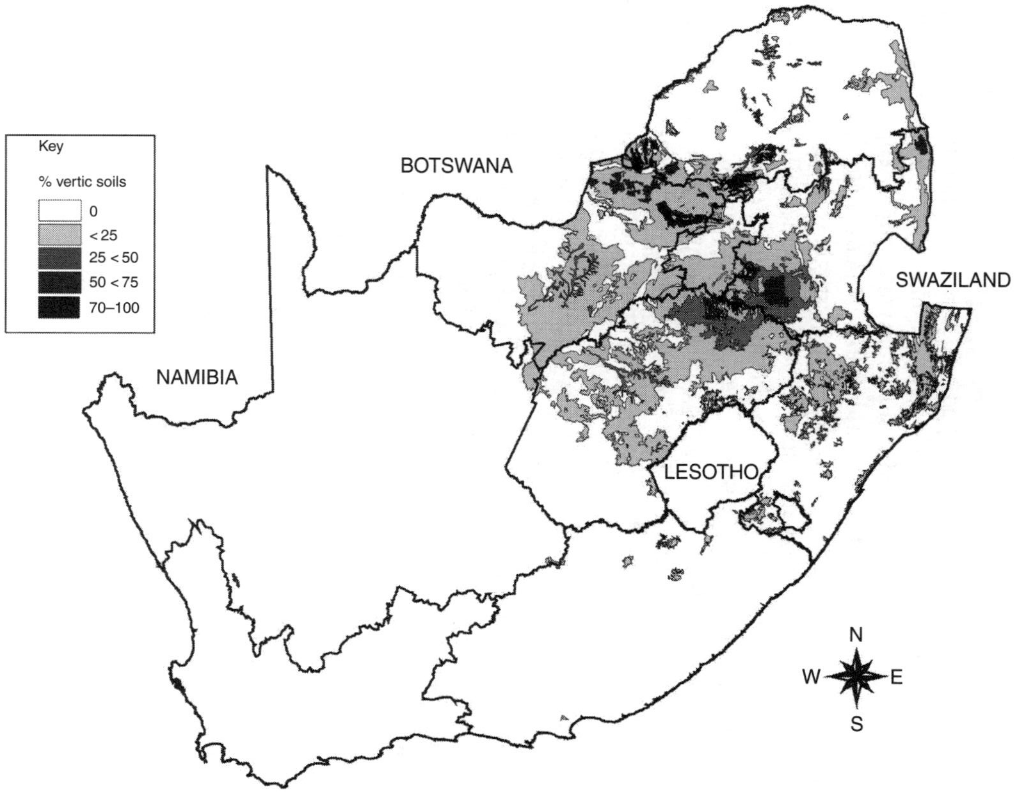

Fig. 6.1. Spatial distribution of Vertisols in South Africa.

The Towoomba Research Station, situated within the Springbok Flats, has a mean annual precipitation of 588.3 mm, though rainfall figures range from 517 to 1152 mm. The mean potential evaporation rate is 6.9 mm day^{-1} and the mean annual maximum air temperature is 26.9°C. Erasmus (1973) recorded as much as 60% evaporation for this station during the summer months (October to March). Figure 6.2 depicts the dominant climatic regimes of South Africa.

Basalt can, for all practical reasons, be regarded as an aquifer for the Springbok Flats. The basalt is cracked and seamed, resembling a heterogeneous sponge rather than an underground lake. Boreholes sunk in these seams and cracks have a high and sustainable water supply capacity, but the apparent absence of separate soil water compartments could be a threat to the whole area as all boreholes are fed by this single source. From 1961 to 1965, water levels declined significantly and supplementation by rain was insufficient to restore water levels to those experienced before monitoring (Department of Water Affairs, 1996).

South Africa with its peculiar geology developed its own soil classification system (Soil Classification Working Group, 1991). Vertisols are classified as Arcadia and Rensburg soil forms, which correlate with Uderts, Usterts and Aquerts of the US/International Soil Taxonomy (Bühmann and Schoeman, 1995). Two Vertisol bodies with significant structural differences are encountered in South Africa, their distinctive features being either self-mulching or crusting. The self-mulching group is derived predominantly from basic igneous rocks and is normally dark grey to black while the more lightly coloured and crusting counterparts are developed, mainly from shales and mudstones of the Karoo Supergroup (Snyman *et al.*, 1985; Bühmann, 1995; Bühmann and Schoeman, 1995). This parent material–surface morphology relationship is consistent with reports from other countries and continents.

Discrete smectite is the dominant clay component in mafic igneous rock-derived Vertisols with kaolinite a common accessory phase while shale-derived counterparts are often associated with random illite/smectite interstratifications (Bühmann and Schoeman, 1995). Crusting and self-mulching soils share a common mineralogy within broad ranges, however, with no consistent differences in the layer-charge characteristics of the swelling clay. The percentage of exchangeable sodium, generally accepted to promote a high degree of dispersion and swelling and accordingly crust formation, cannot reliably predict Vertisol surface morphology (Bühmann, 1995). In South Africa, soils developed from mafic rocks tend to develop crusts after prolonged periods of cultivation (Bühmann, 1995).

Economic importance

Major food crops grown on South African Vertisols are grain sorghum, sunflower, wheat and maize. Cotton and tobacco represent other economically

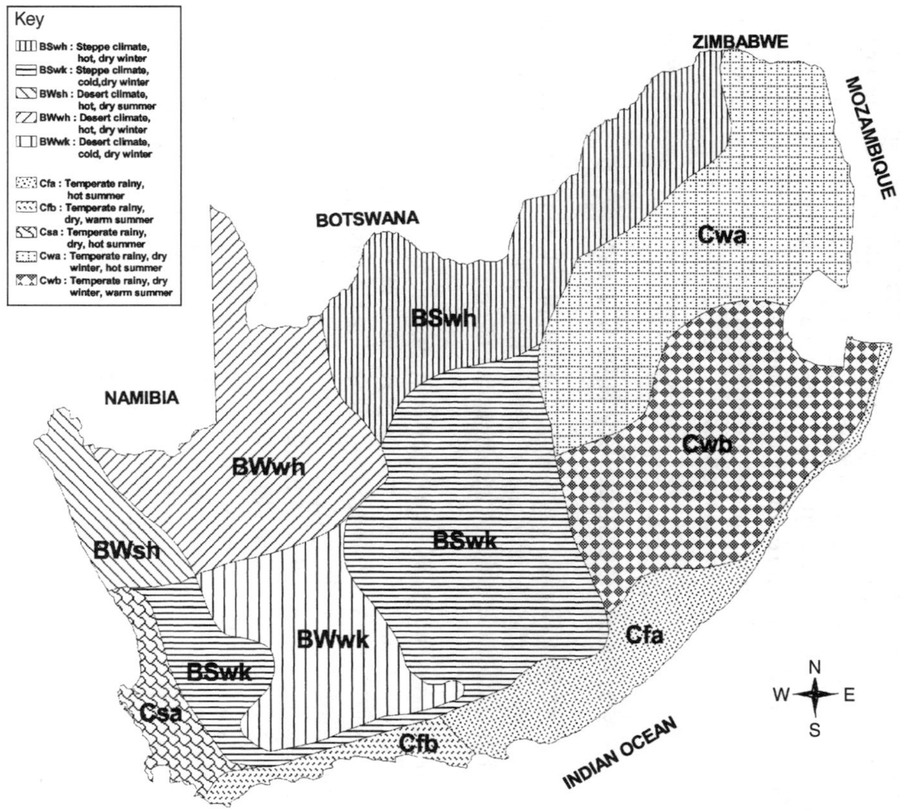

Fig. 6.2. Simplified Köppen climate classification based on rainfall, evaporation and temperature. (ARC/LNR, Institute for Soil, Climate and Water.)

important crops. Most farming systems include an animal factor (Jacobs and Beukes, 1997).

South Africa's Vertisols are relatively productive. Agricultural production on the Springbok Flats, for example, which comprises 5% of the country's arable Vertisols and 1.1% of the country's total arable land clearly illustrates the economic importance of Vertisols in South Africa. According to Grubb (1993), the Springbok Flats accounted for 7.4% of the country's sunflower production, 17.4% of its cotton, 6.2% of the sorghum, 3% of the wheat and 1.3% of the maize production. These yields were obtained from an estimated 10,000 ha on farms with an average size of 400–500 ha, using intensive rainfed production systems (Coetzee, 1993).

In the Springbok Flats about 70% of the area is used for crop production while 30% is used as pasture. Planted pastures are difficult to establish but *Cenchrus ciliaris* in particular is increasingly planted as fallow (D. Coetzee, Towoomba, personal communication).

Vertisol Productivity and Constraints

Agricultural productivity is firmly rooted in the physicochemical–biological properties of the soil. Soil is an open system within an ecosystem and strongly influenced by the external environment with emphasis on management practices.

South African Vertisols are relatively productive with a neutral to slightly acid pH (Grubb, 1993) and display distinct physical and chemical characteristics. These Vertisols are typically prone to crusting and subsequent runoff and erosion, a decline in soil organic matter (OM) content following long-term cultivation, impeded water infiltration and poor availability of water in the plant root zone; they are also prone to waterlogging and anaerobic soil conditions, which could cause an estimated 30% drop in crop yield (Jacobs and Beukes, 1997). According to Erasmus (1973) crops grown on Vertisols can experience drought stress at a moisture content of 20–25% and up to 80% of the precipitation during a high-intensity rainstorm can be lost as a result of runoff. Vertisols have a particularly high soil water retention capacity, a feature typical of clay soils, but low plant-available water is the single largest constraint to crop production.

The behaviour of Vertisols under wet and dry conditions and the extent to which they shrink and swell influence the degree of surface structural phenomena such as self-mulching, crusting, and even the distribution of gilgai mounds and zones affected by crater disease, caused by the pathogen *Rhizoctonia solani* (Grubb, 1993). This change in soil structure is undoubtedly the most dominant threat to soil productivity and crop yield potential. Soil structure deterioration and the consequent deterioration of pores disturbs the natural equilibrium and affects nutrient balance, soil water availability, soil chemistry and soil biological activity. A complicating factor is that structural degradation in Vertisols as a result of cultivation varies both on a micro- and macro-structure scale (Grubb, 1993). The management of Vertisols in South Africa and sustainable cultivation practices are therefore more complicated than those for other soil types and of particular importance if the productivity of this important resource is to be maintained.

The effect of cultivation and the physical, chemical and biological constraints of South African Vertisols are discussed.

Physical constraints

Despite their pedological similarities, small but significant differences in structure have been identified among Vertisols and discernible changes in soil structure are induced by cultivation (Grubb, 1993). Beukes (1987, 1992), in a long-term trial, evaluated the effects of no-till, mouldboard and disc cultivation on the physical properties of Vertisols and on grain sorghum growth and yield. The no-till and tine tillage practices with stubble retained resulted in the highest soil water retention and a significant increase in water infiltration and grain yield, particularly during dry spells (Beukes *et al.*, 1998). The same cultivation practices but with stubble removed, in contrast, resulted in the lowest water retention (Jacobs and Beukes, 1997). This is in accordance with the results of Opperman and Barnard (1982) and Smith and Wehner (1987), namely that, even though soil moisture is affected differently by using different conventional tillage methods, all such practices are unsustainable and result in deteriorating physical and biological properties. It also substantiates Fischer's (1982) conclusion that intensive cultivation of Vertisols and the absence of stubble contribute little, if anything, towards improving soil water availability and maize yield. An effective mulch cover, according to Fischer (1982), not only enhances plant water availability but is conducive to soil structure stability and to mitigating the impact of raindrops. Humic acids from stubble are furthermore adsorbed on to soil minerals and significantly increase aggregate stability, which persists with time (Chaney and Swift, 1986).

The most common cultivation practice in the Springbok Flats is shallow ploughing, as it is the least problematic and cost-intensive. This practice, however, is highly conducive to structural degradation and subsequent compaction. In the Springbok Flats, soil compaction was detected to a depth of 300 mm starting with layer compaction, which ultimately became solid compaction. This is attributed particularly to Na dispersion/deflocculation. The stubble mulch practice is gaining popularity particularly in rainfed sunflower–grain sorghum–cotton rotations where the cotton is slashed (D. Coetzee, Towoomba, 1999, personal communication).

Chemical constraints

The sustainability of any crop production system depends on maintaining adequate soil plant nutrients, organic carbon (C) levels and a favourable C : N ratio. Vertisols are perceived to remain fertile even without the application of fertilizer (Grubb, 1993) despite inherent and widespread low nitrogen (N) and phosphorus (P) levels (Katyal *et al.*, 1987) and potassium (K) fixation (Steenkamp, 1963). On Springbok Flats Vertisols with a clay content of > 50%, no fertilizers are applied despite long-term cultivation while N is applied at intervals on Vertisols with < 30% clay (D. Coetzee, Towoomba, 1999, personal communication). According to Tredoux (1993) the C : N ratio of Springbok Flats Vertisols is low enough to ensure that N is freely available for autotrophic nitrifying bacteria.

Long-term and intensive cultivation of Vertisols, however, affects productivity. Soil organic matter is the major source of N in Vertisols, but if the total carbon content is reduced to < 1% as a result of cultivation, the N-rendering capability and ameliorating effect of OM could be suppressed (Jacobs and Beukes, 1997). Girma (1998) established that cultivation reduces OM levels by as much as 15% at a 0–5 cm depth, causing greater oxidation. This is supported by Torbert *et al.* (1998), who examined the impact of tillage intensity and crop residue on C and N cycling in a Vertisol. They established that chisel tillage reduced total N, C mineralization and C turnover at the 0–10 cm depth compared with the other conservation tillage systems. They concluded that, in the short term, tillage systems may control soil organic C at the soil surface while changes in plant rooting may control organic C at deeper soil depths. Torbert *et al.* (1998) further established that the greater N mineralization occurred with the no-till system, indicating that the reduction in C mineralization and C turnover observed with the conservation tillage systems was not due to N limitations.

High nitrate concentrations are reported for some parts of the Springbok Flats. Soil cultivation was obviously the sole process that caused mobilization and leaching of nitrate followed by accumulation in the sub-surface beyond the root zone (Tredoux, 1993). This finding was confirmed by means of a combined isotopic ($^{15}N/^{14}N$) and chemical study, which related elevated nitrate levels in groundwater directly to cultivation.

Beukes (1992) reported on the long-term effects of stubble and conventional tillage on aggregate stability and other Vertisol properties following a 6-year cultivation trial with grain sorghum in the highveld central parts of South Africa. The area comprises 543,000 ha of Vertisols planted to mainly grain sorghum (85%) and sunflower (8%). This area is characterized by low yield and even crop failure due to low and erratic rainfall. Beukes concluded that a tillage system retaining most of the stubble on the soil surface maintains and increases aggregate stability markedly, as well as soil C and N, while the conventional mouldboard/disc practice resulted in a decrease in these properties and a resultant decrease in soil productivity. Beukes (1992) also established that periodic tine cultivation to control weeds did not seriously affect the soil C level.

In a further study in the same highveld Vertisol area on the effects of N and P fertilizer application on soil properties and grain sorghum yield, Beukes (1989) established that N fertilization had no immediate effect on yield but that the subsequent increase in inorganic soil N secured yield for the following season.

For more than 50 years the most frequently used cultivation practices on the Springbok Flats have been deep ripping, disc ploughing and sequential rotation with sorghum, sunflower and wheat. There is, however, unfortunately no systematic monitoring of soil quality indicators. On the farms being cultivated since 1947, chemical change is reflected predominantly in an increase in pH, which is most pronounced in gilgai mounds. Grubb (1993) reported a pH (KCl) of 7.00 in a gilgai mound compared with 6.00 for the shelf zone. Grubb (1993) also reported a predictable reduction in C, K, N and P during a 7-year field and experimental trial.

Steenkamp (1963) studied the response of tobacco to K applications in a 10-year NPK trial, as K is of particular importance to the quality of tobacco. He concluded that both the exchangeable K status and the capacity of Vertisols to supply K are low due to K fixation. This he ascribed to the specific mineralogy of Vertisols, which is dominated by members of the smectite group. The species involved is mostly beidellite (Bühmann and Schoeman, 1995), i.e. a swelling clay mineral with a relatively high layer charge, arising predominantly from tetrahedral substitutions. The interlayer charge density of beidellite is in many cases high enough to fix K – and to a lesser degree also N – in the interlayer regions, thus rendering it available to plants only under stress conditions. The fixation of K in the interlayers results in a micaceous structure and resultant decrease in the cation exchange capacity of the soil.

Due to the large degree of variability of many Vertisol parameters, reduced crop yields, particularly those associated with gilgai mounds and crater disease zones in the Springbok Flats, are not necessarily associated with nutrient deficiencies. Grubb (1993) compared plant analyses of sunflower and wheat grown on gilgai mounds, crater and shelf zones and recorded higher nutrient levels on gilgai mounds and crater zones than on associated shelf zones. Nutrients compared were P, N, Mn, Cu, Ca, Zn and Fe. Grubb (1993) concluded that the low and erratic rainfall pattern virtually negates the influence of nutrient status in favour of water availability. This is substantiated by increased crop yields when irrigated or in the case of good seasonal rainfall.

South African Vertisols are found in the summer rainfall zones and, as agronomic rather than horticultural crops are grown, irrigation, depending on the availability of water resources, forms part of farming systems for winter crops only. The major problem with irrigated land is compaction, evidently related to increased exchangeable Na levels. Salinization in the Springbok Flats, contrary to irrigated Vertisols in Australia, is believed to be a consequence of prevailing rainfed conditions (Grubb, 1993).

Biological constraints

Soil macro-invertebrates, and especially earthworms and termites, play a fundamental role in the dynamics of soil organic matter, structure and pore distribution at different scales of time and space. These invertebrates furthermore play an important role as disseminators and activators of soil microorganisms, especially fungal spores, which they transport on their bodies, and/or spread in their faecal pellets. Arthropod populations are concentrated generally in the top 5 cm of Vertisols, and burrowing earthworm species improve drainage to a great extent (Loranger *et al.*, 1998), thus countering the waterlogging to which Vertisols are often subjected.

Vertisols are notorious for variable crop response and crop disease directly related to soil structure and water availability (Lubbe, 1988; Grubb, 1993). The integrated management of Vertisols is therefore of critical importance. Maize was the most profitable field crop grown in the Springbok Flats for decades but in the late 1960s wheat was cultivated increasingly. The monoculture wheat production system consisted of bare fallow between successive crops. After a decade, summer wheat yields declined significantly because of crater disease. The yield decline was ascribed to fungi-associated seedling blight, stunting and premature death of plants in patches. Scott *et al.* (1979), in a disease study, isolated *Periconia macrospinosa*, *Pythium oligandrum*, *Rhizoctonia cerealis* and *Rhizoctonia solani*, which all formed characteristic microsclerotia, resting spores or mycelium in the root cortex of infected plants. These authors concluded that crater disease is reminiscent of root diseases caused by phytotoxins rather than those caused by root-rotting fungi. Further crater disease studies in the Springbok Flats concentrated largely on *R. solani*. This disease resembles bare patch disease of cereals occurring in Australia, in Scotland and in parts of America. Crater disease was the most serious constraint to wheat production in the Springbok Flats, where an estimated 60,000 ha was planted to wheat annually (Maas and Kotze, 1981; Lubbe, 1988) and Smith and Wehner (1987) calculated that crater disease was responsible for wheat crop losses of up to 35%. Wheat production in the Springbok Flats has since been reduced drastically.

Various tillage treatments influence the incidence of crater disease to a limited extent (Lubbe, 1988; Opperman and Barnard, 1992), while a significant decrease in the incidence and viability of *R. solani* in crater disease was achieved through crop rotation or fallow compared with intensive wheat cultivation (Lubbe, 1988). The crops rotated were sunflower, maize, sorghum, soybeans and cotton.

Lubbe (1988) found that *R. solani* in crater disease differed from other strains of the organism in having a considerably lower growth rate and in being unable to use nitrite as a sole source of nitrogen. *R. solani* produced compounds phytotoxic to wheat *in vitro*. This was confirmed by Meyer (1996), who found that *R. solani* represented a distinct intraspecific group within the anastomosis group 6 of the species. Restriction fragment length polymorphism studies established the presence of a unique 610 bp fragment in 25 rDNA of *R. solani*. While *R. solani* could not be isolated from native grasses it was associated with the umbrella-thorn tree, *Acacia tortilis* subsp. *heteracantha*, and also with patchy stunting of wheat on Vertisols in Tanzania. *R. solani* was also distinct from the strain causing bare patch disease in Australia, which belongs to anastomosis group 8. The crater disease example amply illustrates the fact that Vertisols can exhibit unique phenomena with regard to disease when compared with lighter soils and require special attention when diagnosing causes.

The preponderance of pest groups such as nematodes on lighter soils is well known, while clay soils with high levels of smectite, such as Vertisols, are relatively free of this scourge. In the case of insect pests, soil texture and colour may affect the distribution of particular species. On the white sandy soils of the Cape Flats the white cutworm, *Euxoa subalba*, predominates, while the common cutworm, *Agrotis segetum*, and the black cutworm, *Agrotis ipsilon*, predominate on heavier soils and Vertisols. A particular case where an insect pest is relatively minor on lighter soils but can be responsible for considerable damage to maize on Vertisols is documented by Walters (1979) in the case of the spotted maize beetle, *Astylus atromaculatus*. In the case of Vertisols the seed is sown under dry conditions as the fields are inaccessible after rains. Losses to the pest larvae often necessitate replanting of the fields. In South Africa, maize is produced mainly on the red and yellow lateritic soils but it can be expected that distinct differences between the pest and disease complexes will become further evident as the use of Vertisols increases, perhaps as a result of greater implementation of no-tillage practices. The dominant threat to monoculture maize production is the stemborer. In the case of the African stemborer, *Busseola fusca*, Walters (1975) inferred that the emergence of moths from buried crop residues from the previous season was dependent upon the existence of open channels allowing egress to the surface. In the case of Vertisols that are dry-ploughed, residues are ineffectively buried and emergence of moths facilitated. More intensive use of Vertisols from crop production, and even that of cultivated pastures, will therefore require integrated pest and disease management systems that will differ from those required on other soil types. The introduction of conservation and no-tillage practices will facilitate access under moist conditions, which would be difficult under normal cultivation practices. This will affect the damage by and occurrence of pests, while the presence of disease inocula on crop residues will place greater emphasis on the management of crop rotation and the development of resistant cultivars.

Adoption

The technology adoption process is dynamic as it involves the characteristics of both the user and the technology (Düvel, 1997; Jabbar *et al.*, 1998). Non-adoption can be traced back to either the incapability or unwillingness of the individual to adopt (Düvel, 1997). Unwillingness is associated usually with the lack of immediate incentive or with a variety of socio-economic aspects. As the input cost to produce food on Vertisols is comparatively low, the commercial farmers of the Springbok Flats are quite satisfied with a favourable profit margin gained from sub-optimal wheat yields of 0.5–0.75 t ha^{-1} $year^{-1}$ (D. Coetzee, Towoomba, 1999, personal communication). This is despite technology transfer on long-term research on the sustainable management of Vertisols in the Springbok Flats. The adoption rate, according to extensionists, is low and soil structure degradation and soil compaction are perceived to increase, although they are not monitored systematically. Bearing in mind South Africa's population growth rate of 2.5% per annum, it is imperative that the sustainable output on Vertisols, requiring relatively low input for food production, be expanded.

The authors endorse Düvel's (1997) belief that adoption of technology can best be facilitated by on-farm and demonstration trials where the farmers, researchers, extensionists, advisers and farmer-researchers operate as a team. This approach is substantiated by International Crops Research Institute for the Semi-Arid Tropics (ICRISAT) (Von Oppen *et al.*, 1989). ICRISAT's improved Vertisol technology was developed from the component research stage to the package-and-system design stage on the research station and then introduced into farmers' fields. The constraints were understood only when farmers were confronted with the technology. It is envisaged that smallholder settlements on Vertisols in South Africa will realize a higher adoption rate of sustainable farming systems and practices if greater emphasis is placed on on-farm research, demonstration trials and small agribusiness development, which is an incentive. Towoomba Research Station staff established that one of the communities within a certain district planting maize only has regular crop failure, while a neighbouring community practising crop rotation with grain sorghum and millet do not experience crop failures. It is envisaged that the community practising monoculture production will gradually adopt sustainable farming systems but to change cultural practices is a long-term endeavour requiring insight into prevailing systems, beliefs and incentives.

Few, if any, real-time statistics on land use, total number of households, farming practices, yield and farm animals are available for Vertisols. This is particularly true for the Springbok Flats. The Northern Province Department of Agriculture, responsible for the Springbok Flats, identified the establishment of an organized agricultural union catering for both commercial and small-scale farmers as an urgent requirement. The KwaZulu-Natal Province is far advanced in this regard and community-based marketing organizations have been established. The resettlement of small-scale farmers and communities on previously State-owned land in the Springbok Flats and land claims by the previously disadvantaged are indisputable facts necessitating such an organization to ensure sustainable land-use planning and goal-directed services on Vertisols, which are difficult to manage. It should furthermore be emphasized that the basic requirement of resource-poor communities being resettled is household food security, but it is envisaged that some community members will develop into fully fledged farmers. The adoption of sustainable Vertisol management practices is therefore of great importance. In a Springbok Flats district where a community was recently resettled on 3000 m^2, 14 households, each with 0.3 ha land, plant vegetables for subsistence (Greeff, Potgietersrust, 1999, personal communication). The major constraint obviously is plant-available soil water. Boreholes were sunk and equipped to enable the watering of vegetable gardens but, considering the geohydrology of the Springbok Flats, the water resource could be depleted if the Vertisols are not managed sustainably and in an integrated manner.

Donaldson (Towoomba, 1999, personal communication) is presently completing a study on the control of maize stemborer. Sweet sorghum (*Sorghum bicolor*), maize and selected grass species are rotated. The incidence of stemborers has declined significantly while the rate of alkalinization is curbed by the sweet sorghum cultivated under irrigation. A major advantage of this research is that an extractor suitable for the smallholder was designed to extract molasses from the sweet sorghum. This additional incentive to change agricultural practices is, according to the researcher, received with enthusiasm as it provides the opportunity for much needed small business development.

The Institute for Soil, Climate and Water, in cooperation with the Provincial Department of Agriculture concerned, evaluates water harvesting and basin tillage techniques on heavy clay soils under low and erratic rainfall. Crops rotated are maize, sunflower, wheat and grain sorghum. The evaluation is both experimental and via on-farm demonstrations, with a subsequent high adoption rate. Field crop yield increased significantly within three seasons.

Future Considerations in Vertisol Management

To arrest the degradation of Vertisol soil structure and subsequently crop failure through water stress, soil erosion losses and disease patterns (Grubb, 1993), the following integrated agrobiological practices should be further researched and adoption facilitated by on-farm trials supported by farmer research:

- residue management by retaining sufficient crop residue;
- no-tillage practices with tine and disc ploughing when necessary to promote carbon sequestration in the upper layers of the soil;
- crop rotation patterns;
- alternating irrigated crop production (winter wheat) with rainfed cropping;
- cotton cultivation on weakly aggregated Vertisols showing severe gilgai-related crop losses;
- strip cultivation;
- maintenance of active soil biological processes; and

- comprehensive studies to elucidate the vagaries of sustainable Vertisol management.

It should be borne in mind that Vertisols are both complex and varied and that management practices should be in accordance with the inherently variable Vertisol characteristics.

Conclusion

Because Vertisol behaviour is both complex and varied, no single solution can be offered for their effective management. An integrated soil management system is therefore recommended as it takes charge of soil physicochemical–biological properties within an open system influenced by climate, socio-cultural, economic, environmental and management factors. With the integrated management of Vertisols, latent and apparent structural change may be moderated or even modified to the advantage of sustainability and the land user.

Vertisols are inherently productive and should be used more intensively and extensively for field crop production as production costs can be comparatively low if the soil physicochemical–biological constraints are effectively managed. This is of particular importance in South Africa as high-density resource-poor communities are resident on Vertisols.

The major Vertisol constraint to successful field production is plant water availability. It has been proved experimentally as well as by means of on-farm trials that soil water availability at the plant root level can be improved significantly by retaining stubble and conservation tillage favourable to soil biological activity and disease control. Water harvesting is a further solution to efficient water use. The integrated management of Vertisols is furthermore conducive to soil structure stability, which is of the utmost importance to the sustainable management of Vertisols.

The prerequisite for the successful management of Vertisols is, however, that the sustainable management practices proposed are adopted and that technology problems are referred back to the research component. The technology transfer cycle can be facilitated most effectively by on-farm research.

References

Beukes, D.J. (1987) The effects of different stubble-mulch systems on certain soil properties as well as on growth and yield of grain sorghum on a vertic soil. Paper presented at 14th Soil Science Society South Africa Congress, Nelspruit, South Africa.

Beukes, D.J. (1989) The effects of N and P fertilizer on soil properties and on yield of grain sorghum on a vertic soil. Paper presented at Combined Soil Science Society South Africa and Crop Production Congress, Wild Coast, South Africa.

Beukes, D.J. (1992) Long-term effects of stubble and conventional tillage on aggregate stability and other properties of a vertic soil. Paper presented at Soil Science Society South Africa Congress, Stellenbosch, South Africa.

Beukes, D.J., Bennie, A.T.P. and Hensley, M. (1998) Optimization of soil water use in the dry crop production areas of South Africa. In: van Duivenbooden, N., Pala, M., Studer, C. and Bielders, C.L. (eds) *Proceedings of OSWO Consortium Workshop on Efficient Soil Water Use: the Key to Sustainable Crop Production in Dry Areas of West Asia, and North and Sub-Saharan Africa.* ICRISAT, Niamey, Niger, pp. 165–191.

Bühmann, C. (1995) Mineralogical and chemical characteristics of crusting and self-mulching vertisols from South Africa. In: *Proceedings ARC-ISCW Wise Land Use Symposium*. ISCW, Pretoria, South Africa, pp. 54–59.

Bühmann, C. and Schoeman, J.L. (1995) A mineralogical characterization of vertisols from the northern regions of the Republic of South Africa. *Geoderma* 66, 239–257.

Chaney, K. and Swift, R.S. (1986) Studies on aggregate stability. II. The effect of humic substances on the stability of re-formed soil aggregates. *Journal of Soil Science* 37, 337–343.

Coetzee, D. (1993) Leer turf eers ken. *Agricultural News* 5 April, 5.

Department of Water Affairs (1996) Verslag oor grondwateraangeleenthede op die Springbokvlakte. *Department of Water Affairs Technical Report* No. GH 3056, Pretoria, South Africa.

Düvel, G.H. (1997) The human factor in sustainable agricultural land use. In: *Proceedings Soil Acidity Initiative (venue Mpumalanga) Symposium*, Ermelo, South Africa, pp. 45–50.

Erasmus, R. (1973) Vogopgaring, vogbewaring en vogbenutting op die Springbokvlakte. Seminar, University of Pretoria, South Africa.

Fischer, H.H. (1982) Grondbewerkings- en vogbewaringspraktyke op die swartkleigronde van die Springbokvlakte. *Crop Production* XI, 81–84.

Girma, T. (1998) Effect of cultivation on physical and chemical properties of a Vertisol in Middle Awash Valley, Ethiopia. *Communication in Soil Science and Plant Analysis* 29(586), 587–589.

Grubb, P.L.C. (1993) Vertisols in the Springbok Flats, Northern Transvaal: some aspects of their nature, cultivation and structural degradation. *ARC-ISCW Report* No. GW/A/94/32, Pretoria, South Africa.

Jabbar, M.A., Beyene, H., Saleem, M.A.M. and Gebreselassie, S. (1998) Adoption pathways for new agricultural technologies: an approach and an application to Vertisol management technology in Ethiopia. *Socioeconomic and Policy Research Working Paper* No. 23, Ethiopia.

Jacobs, E.O. and Beukes, D.J. (1997) The nature and sustainable management of vertisols in South Africa. Paper presented to the SARCCUS Standing Committee for Soil Science, Pretoria, South Africa.

Katyal, J.C., Hong, C.W. and Vlek, P.L.G. (1987) Fertilizer management in Vertisols. In: Management of Vertisols under Semi-arid Conditions. IBSRAM Proceedings No. 6, IBSRAM, Bankok, Thailand, pp. 247–266.

Loranger, G., Ponge, J.F., Blamchart, E. and Lavelle, P. (1998) Impact of earthworms on the diversity of microarthropods in a Vertisol (Martinique). *Biology and Fertility of Soils* 27, 21–26.

Lubbe, E.M. (1988) Etiology and control of crater disease of wheat. DSc (Agric.) thesis, University Pretoria, South Africa.

Maas, E.M.C. and Kotze, J.M. (1981) Research note. Fungi associated with root disease of wheat in South Africa. *Phytophylactica* 13, 155–156.

Meyer, L. (1996) Characteristics and ecology of *Rhizoctonia solani* causing crater disease of wheat. PhD thesis, University of Pretoria, South Africa.

Opperman, L. and Barnard, R.O. (1982) Improved growth and grain yield of wheat in deep mouldboard ploughed crater disease and disease-free soil on the Springbok Flats. *Phytophylactica* 24, 335–338.

Scott, D.B., Visser, C.P.N. and Rufenacht, E.M.C. (1979) Crater disease of summer wheat in African drylands. *Plant Disease Reporter* 83, 836–840.

Smith, E. and Wehner, F.E. (1987) Biological and chemical measures integrated with deep soil cultivation against crater disease of wheat. *Phytophylactica* 19, 87–90.

Snyman, K., Fey, M.V. and Cass, A. (1985) Physical properties of some highveld Vertisols. *South African Journal of Plant Soil* 2, 18–20.

Soil Classification Working Group (1991) Soil classification: a taxonomic system for South Africa. *Memoir Agricultural Natural Resources South Africa* No. 15. Department Agricultural Development, Pretoria, South Africa.

Steenkamp, C.J. (1963) Kalium ewewigte in die swart turfgrond en die opname daarvan deur tabak. MSc (Agric.) thesis, University Stellenbosch, South Africa.

Torbert, H.A., Potter, K.N. and Morrison, J.E. (1998) Tillage intensity and crop residue effects on nitrogen and carbon cycling in a Vertisol. *Communications in Soil Science and Plant Analysis* 29(5 & 6), 717–727.

Tredoux, G. (1993) A preliminary investigation of the nitrate content of groundwater and limitation of nitrate input. *Water Research Commission (WRC) Report* No. 368/1/93, Pretoria, South Africa.

van der Merwe, A.J. (1995) Wise land use: the basis for sustainable growth and development in South Africa. In: *Proceedings ARC-ISCW Wise Land Use Symposium*. Pretoria, South Africa, pp. 2–8.

Von Oppen, M., Ghodake, R.D., Kshirsagar, K.G. and Singh, R.P. (1989) Socioeconomic aspects of transfer of technology. In: *Proceedings IBSRAM Inaugural Workshop, India*. IBSRAM, Bangkok, pp. 235–249.

Walters, M.C. (1975) Evolution in tillage techniques and impact on entomological research, with special reference to the maize stalkborer, *Busseola fusca* (Fuller). In: *Proceedings 1st Congress of the Entomological Society of Southern Africa 1974*. Entomological Society South Africa, Pretoria pp., 235–244.

Walters, M.C. (1979) Maize pests of sub-Saharan Africa. In: Häflinger, E.H. (ed.) *Maize, Ciba-Geigy Agrochemicals Technical Monograph*. Ciba Geigy, Basel, Switzerland, pp. 66–77.

Vertisols Management in the Sudan

7

A.T. AYOUB

Khartoum, Sudan

Key words: Sudan, Vertisols, soil management, soil physical properties, fertilizers, soil degradation

Introduction

The Sudan has a vast expanse of Vertisol plains, which are considered the most important soils. Vertisols cover an area of about 70 million ha, and about 28% of the country's total area (FAO, 1995). These soils comprise the central, eastern, western and southern Vertisols. The central Vertisol stretches from south of Khartoum between the Blue and White Niles southwards to Sobat River at the fringe of southern Sudan. It covers an area of about 21 million ha. The eastern Vertisol is between the Blue Nile and Atbara Rivers towards the borders with Ethiopia and covers an area of about 16 million ha. The western Vertisol is west of the White Nile in Southern Kordofan State, totalling about 12 million ha. The southern Vertisol is in southern Sudan, east and west of the flood plain, and covers an area of about 21 million ha. Seventy per cent of the Sudan Vertisols are in the northern part of the country and the rest are in the south. Approximately 67% of the Sudan's Vertisols receive dependable rainfall (between 600 and 800 mm year^{-1}). The portion receiving unreliable rainfall is exclusively in the north of the country. In general terms, Vertisols in the north are more fertile and easily workable than those in the south.

The parent material of the central Vertisol is the Blue Nile alluvium from basaltic rocks in Ethiopia laid down when the area was a lake from 50,000 to 10,000 BC. Those of the eastern and western Vertisols are of basaltic origin, weathered *in situ*, and the southern Vertisol is believed to have been deposited by the White Nile (Tothill, 1954).

In the last 30 years, the Sudan has expanded its agricultural production mainly through an increase in rainfed cropland area, which stands today at about 20 million ha; 42% is under large-scale mechanized agriculture and the remainder is under small-scale subsistence traditional cultivation. Soon after the introduction of large-scale mechanized rainfed agriculture, conflicting goals between such increased production and sustainable resource management became evident. The purpose of this chapter is to summarize the work that has been done in the Sudan to improve the agricultural management of Vertisols.

The Vertisols Plain

The Vertisols plain lies in the arid and semi-arid zones, with an erratic mean annual rainfall varying from 400 to 800 mm, and a general elevation of 350–450 m above sea level. The natural habitat in the north is composed mainly of tall grass plus *Acacia* woodland while in the south the area is inundated seasonally, forming part of the flood plain with vast meadows of almost pure grass. Apart from the irrigated areas, the Vertisols plain is distinguished as being the zone of both animal and rainfed husbandry.

General Characteristics of the Sudan Vertisols

The Sudan Vertisols have developed under varying environments, but have the following general characteristics in common:

1. Clay content of more than 40% throughout the upper 1 to 2 m; with a clay to cation exchange capacity ratio of 1 : 1.
2. Wedge-shaped parallelepiped aggregates.
3. Dominance of montmorillonite mineral in the clay fraction.
4. Cracks at least 2 cm wide and over 50 cm deep develop in most environments.
5. Low organic matter content ranging between 0.5 and 1.5%.
6. Very slow permeability when wet.
7. The profile is free from gravel and stones – fractions larger than 2 mm are dark grey calcium carbonate concretions.
8. B horizon hard to recognize.

One of the extensive and most representative soil series in the irrigated Vertisols of the central clay plain is the Suleimi series (El Tom, 1987). These soils are Entic Chromusterts, members of the fine montmorillonitic isohyperthermic family occupying level parts of the central Vertisols. Most of the research studies at the Gezira Research Farm have been conducted on these soil series. The soil profile is deep, down to 10 m, and moderately well drained because of the deep cracks which remain open over six cumulative months per year. Gilgai relief also develops, providing a temporary physical improvement through self-mulching. The soil has a clay content of 55 to 60% and moderate to strong alkalinity. As for the rainfed agriculture, the Dinder series is the most extensive, occupying large areas of the southern part of the central Vertisol and most of the eastern Vertisol (El Tom, 1987). These are Typic Chromusterts of a very fine montmorillonitic isohyperthermic family occurring on level to nearly level alluvial deposits. Like the Suleimi series, the Dinder series is deep and has a very dark brown surface with soft mulch and a greyish to yellowish grey sub-surface. Clay minerals are dominated by montmorillonite with some kaolinite in both the Suleimi and Dinder series. Tillage is difficult in both soil series unless the soil has an optimum water content.

Experimental Work on Sustainable Use of the Vertisols

Land preparation

Willcocks (1987) reported research done on land preparation in the rainfed area. His report concluded that, on the self-loosening Vertisols of the central, eastern and western Sudan, a shallow (10 cm) weeding operation was all that was necessary to prepare an adequate seedbed for rainfed agriculture when sufficient rain had fallen to replenish soil moisture through the cracks. This would be done rapidly and economically for the extensive cultivation of the rainfed Vertisols by a high speed, shallow cultivator. Deep tillage of the rainfed agriculture was very costly, relatively slow and produced a cloddy tilth that required further energy to break down the soil to an acceptable seedbed. The energy cost of the chisel system of land preparation for rainfed sorghum per tonne of crop yield was about four times that achieved with shallow tillage and was four times slower than the shallow tillage. Pre-rain tillage covered most of the cracks and inhibited water replenishment of the deeper profile via the cracks. Mahmoud (1988) reported that the use of a disc harrow year after year was the only tillage implement that produced a hard soil layer.

Water relationships

Farbrother (1987) reviewed field trials testing irrigation intervals and rates on the Suleimi soil series from 1925 to 1963. An interval of 14 days between watering, coupled with a 100-mm rate of application, was the optimum irrigation for most of the crops on the irrigated Sudan Vertisols. The cyclic depth of normal irrigation was known to be 60 cm or so; and no treatment such as deep ripping or application of gypsum appeared to have any reproducible effect on the depth of water penetration. Cracks provide the main route for movement of water downwards, and moisture at any specific depth is related to the original cross section of the crack. The available water held by the 0–60 cm depth zone in the post-irrigation state was of the order of 93 mm (929 m^3 ha^{-1}), of which 49%, 33% and 18% were held in intervals of 0–20, 20–40 and 40–60 cm depths, respectively. Farbrother (1987) citing Abdine (1971) reported that a little weight stress depressed the hydraulic conductivity of the Gezira soil to virtually zero, even at low levels of exchangeable sodium percentage.

Crop rotation

Jewitt (1955) reported the results of several long-term rotation experiments carried out at the Gezira Research Farm. A three-course rotation experiment was started in the 1927/28 season and was evaluated after 24 consecutive seasons. A more comprehensive experiment called the Combined Rotations Experiment was started in the 1936/37 season to test the effects of different crops and fallow periods on cotton yield. The first evaluation of the Combined Rotation Experiment was done after 10 years and the results of both experiments showed that cotton yields were increased significantly by increasing the proportion of fallow in the rotation. They also showed that rotations including a legume gave better yields than those with sorghum, particularly when the cotton immediately followed the legume or sorghum. Cotton–legume–fallow gave the highest cotton yield, and cotton–sorghum–legume gave similar yields to cotton–fallow–fallow. Cotton yields were influenced most by the differing physical conditions of the soil in the different rotations. The results of the 14-year rotation experiments showed that cotton yields were also proportional to the average nitrate-nitrogen levels in the soil during August–November, the period of vegetative growth of the cotton (Jewitt, 1955). The rotation cotton–legume–fallow had the highest nitrate-nitrogen of 14 mg kg^{-1}, while the rotation cotton–sorghum–sorghum had a nitrate-nitrogen level of as low as 5 mg kg^{-1}. Rotational cropping did not increase organic nitrogen beyond an equilibrium amount, which was presumed to be a function of environmental conditions, neither did it affect the organic carbon : organic nitrogen ratio (Jones, 1957, 1958).

A four-course rotation cotton–sorghum–legume–fallow was chosen in the irrigated Gezira in the 1950s. Sorghum (the staple food of the tenant), though exhaustive to soil nitrate-nitrogen, was included in the rotation to give security of tenure to the tenant.

Physical and chemical improvements

In these slowly permeable soils, the first test to be done towards soil improvement was to test whether the application of gypsum and other substances would improve soil permeability. Application of gypsum at 5.7 t ha^{-1} increased the amount of water taken by only 5% to about 120 cm depth after 3 days of flooding, while the same amount of gypsum and 14 days of flooding increased the soil moisture content to a depth of about 180 cm by about 19%. Gypsum in coarse crystals or finely ground gave similar results (Jewitt, 1955). Similar results were obtained using equivalent amounts of potassium sulphate, sulphur, sodium sulphate and ferrous sulphate. Corresponding significant increases in cotton yields were obtained in the year of application in plots both fertilized and unfertilized with 407 kg $(NH_4)_2SO_4$ ha^{-1}, but increases were insignificant in later crops.

Some areas of the Vertisols in the Gezira/Managil area known as the Laota series, are high in soluble salts and exchangeable sodium. Trials of several species of saltbush in the 1930s showed that *Atriplex mulleri* removed striking amounts of salts, particularly sodium, from the soil. With the current intensified rotation, a saltbush crop every 7–8 years should be sufficient to remove all sodium added to the soil through irrigation water. The saltbush also removes considerable amounts of nitrogen and potassium from the soil, and its use might well demand the application of additional fertilizers (Jewitt, 1955).

Effects of fertilizers

Many experimental investigations have been undertaken at the Gezira Research Farm since 1919 on the chemical fertility of the Suleimi series Vertisols. Nitrogen deficiency was recognized from the start (Snow, 1948). Later, Burhan and Mansi (1970) reported the results of the effect of a long-term NPK experiment of 18 years on cotton, which confirmed a consistent response of cotton to nitrogen, erratic response to phosphate and negligible response to potassium. Ammonia volatilization caused serious waste of the fertilizer nitrogen (Musa, 1968). Much biological activity was shown in the surface soil, which was stimulated by irrigation or rain, but decreased down the profile (Musa,1970). At the beginning of the season the *Nitrosomonas* population increased more rapidly than *Nitrobacter*, inducing nitrite accumulation for a short time; this was quickly transformed into nitrates, which were leached down the profile through the cracks with the first heavy rains. Ayoub (1986) using labelled urea found that the amount of nitrogen taken up by cotton placed 10–20 cm deep was 45% more than when placed on the surface. Recovery of fertilizer applied to sorghum at sowing time was small, but improved when the fertilizer was applied 4 weeks later. The Gezira soil already has about 0.3 cmol(+) of fixed ammonium kg^{-1} soil, and cropping and fertilizer application had no effect on ammonium fixation (Said, 1973).

Rice responded significantly to phosphorus fertilization. The yield of cotton following rice in the rotation was enhanced significantly by residual phosphorus. Mixing phosphorus with nitrogen increased wheat grain yield by about 30%, and nitrogen recovery by about 50% over the same nitrogen treatment without P (Ayoub, 1986).

The lack of response to potassium fertilization in the Gezira soil led to more investigations. The studies revealed that the soils had only moderate K reserves, but exchangeable K was high (Fink, 1962). Earlier work by Greene and Snow in 1936 showed that significant amounts of potassium were provided to the crop when irrigated by the Blue Nile (Snow, 1948).

Soil Degradation

UNEP/ISRIC (1990) reported that about 3 million ha of the Sudan's eastern Vertisols under mechanized rainfed agriculture in the Gedaref area were experiencing high physical deterioration due to the use of heavy machinery and monocropping. El Tom (1987) reported similar soil physical deterioration in the southern part of the central Vertisol in the Agadi area. Physical deterioration was the least common of the four soil degradation types in the Sudan reported by UNEP/ISRIC (1990), but it could spread if inappropriate management practices continue. The Nuba Mountains in the western Vertisols are experiencing high topsoil loss through water erosion. These soils are vulnerable to erosion due to the sloping terrain denuded of vegetation cover. UNEP/ISRIC (1990) also reported that about 4 million ha of the Vertisols were stable under sustainable agriculture. These mostly include the large irrigated schemes (e.g. Gezira/Managil, New Halfa) and some parts of the rainfed mechanized agriculture. Table 7.1 shows the degradation of major soil types in the Sudan. Soil degradation varies from 16 to 73% for different soil types. The most extensively degraded soils are the Arenosols and Leptasols. Surprisingly, the least degraded are the Vertisols. Only 16% of the total area is reported to be degraded. Vertisols in most cases are on flat terrain and this reduces the chances of sheet and gully erosion. However, under current cropping systems both irrigated and rainfed areas are increasingly confronted with soil physical deterioration, nutrient depletion and loss of soil organic matter.

Current Agricultural Systems and Cropping Patterns

The indispensable role of nitrogen fertilizer in maximizing the yields of cotton, wheat and sorghum on the Sudan Vertisols has been documented. Urea is now used to grow cotton, wheat, sorghum and some horticultural crops on the irrigated Vertisols. Cotton sometimes receives foliar spray of micronutrients and wheat receives phosphates besides nitrogen.

The Gezira and its Managil extension, New Halfa, Rahad, the Blue and the White Nile schemes and the five sugar plantations established mostly on the

Table 7.1. Severity of degradation of major soil types in the Sudan (from Ayoub,1998).

	Soil degradation (% total area)		
Major soil types	Moderate	Severe	Very severe
Vertisols	3	7	9
Ferralsols	15	16	0
Arenosols	3	52	19
Leptosols	0	0	58

Suleimi, Laota and the Dinder series are the main irrigated areas of the Sudan Vertisols comprising an area of about 1.5 million ha. Cotton land is cultivated by a disc harrow at spacing of 80 cm between ridges and 40 cm in the ridge. Land is normally pre-irrigated before sowing if rains do not come at that time. Fourteen waterings are usually given to the cotton crop at about 2-week intervals and these cease at the end of March. Cotton receives 126 kg N ha^{-1} in the form of urea, 84 kg N applied during sowing and 42 kg N applied at re-ridging time. Sorghum and groundnuts are sown in July on ridges 60 cm apart and 20 cm between holes, usually without pre-irrigation. The crops are grown within the rainy season, but normally receive three to four waterings and mature at the end of October. Wheat is sown mechanically on the flat in November and harvested in March and should ideally receive eight waterings. Wheat receives 84 kg N h^{-1} and 18 kg P ha^{-1} at sowing. Land preparation and frequency of irrigation are crucial for crop production on Vertisols.

About 20 million ha of the Vertisols are currently cultivated under rainfed agriculture, 42% under large-scale mechanized agriculture and the remainder under small-scale traditional cultivation (Ministry of Environment and Tourism, 1996). The bulk of the mechanized farming is owned privately on holdings of 420 ha or larger. The small-scale farmer is, on the other hand, a subsistence-oriented producer who meets his production targets on holdings of between 2 and 5 ha. The main crops grown in both systems are sorghum, millet, sesame, groundnuts and cotton. In its early days, the mechanized rainfed agriculture seemed to be successful with good yields. However, as time went by, yields began to decline due to a shortage of appropriate inputs and lack of adequate resource management. For example, rainfed agriculture received hardly any kind of fertilizers.

Mechanized farming is done basically with a four-wheel drive tractor of about 65 hp together with a set of wide level disc harrows fitted with a seed box. Disc cultivation starts when sufficient rain has fallen to partially close the deep cracks in the clay. A major complication is the likely buildup of soil moisture from August to September and tractors are easily bogged down in the wet clay and prevented from working efficiently. Figure 7.1 shows trends in the yields of the crops during 1961–1996. The trend lines show decreases in yields of all crops by about 50% during this period. No explanation of these yield declines can be offered immediately as there are no data from long-term soil monitoring. However, Burford (1987) reviewing an array of research efforts to improve rainfed agriculture on Vertisols in the Sudan, concluded that immediate agronomic problems that could depress yields were: difficulties in soil preparation, sowing outside the optimum time, possibly nutrients, and very severe weed infestations, especially *Striga*. Though the small-scale farmers practise better resource management than the large-scale farmers because they have a longer-term stake in maintaining the resource base, even the small-scale farmers are faced with declining yields because of the shortening of the fallow periods. By the 1970s, about 75% of the small-scale farmers had cultivated the same plot for more than 5 years (Atta El Moula, 1985).

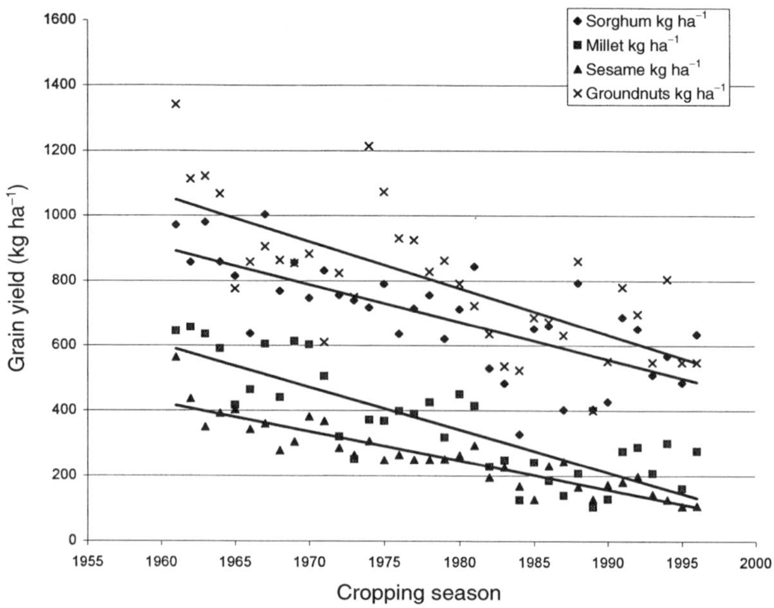

Fig. 7.1. Yields of sorghum, millet, sesame and groundnuts as affected by cropping season. (From Ayoub, 1999. Source: FAO, 1997.)

Discussion

The Sudan has accumulated a respectable record of information on agricultural production technology and management of both irrigated and rainfed Vertisols. Such knowledge has been acquired through longstanding research and management in the irrigated Gezira for about eight decades, and the mechanized rainfed agriculture for about half a century. The Gezira and the rainfed development experiment set very good economic and management examples as models for Sudanese social and economic development. However, lack of appropriate infrastructure, poor economy, political instability and a massive brain drain, which started in 1975, frustrated all planned development (Beshir, 1999). For example, no expansion in irrigated agriculture took place in two decades, and there was a negative impact on the environment by hasty expansion of rainfed agriculture by the private sector, which ignored research recommendations. Government expenditure on research and development has gradually faded away, and for over a decade and a half international assistance in all forms of rehabilitation or development has been totally halted by the donor community.

The sustainability of rainfed agriculture should follow some strategic approaches to maintain soil fertility and conserve moisture. The rainfed agriculture of the Sudan should be classified according to rainfall/evapotranspiration potentials and categorized distinctly according to soil potential. Studies indicate that the amount of moisture stored in the soil and nutrients available are significant measurable factors affecting the production of the crops. Therefore, every effort should be made to store as much moisture as possible in the soil and to reduce evaporation loss to a minimum by means of sound cultural practices. While the incidence of drought may not be reduced, its effects can be mitigated through proper drought management strategies. Mechanized rainfed agriculture in the Sudan is based on the assumption that the wet conditions of the late 1950s and early 1960s are the norm. Such mistaken assumptions are due to misunderstanding of dryland environments and lack of long-term data (Walsh *et al.*, 1988). The tolerable exchangeable sodium percentage in the Gezira soil was up to 25% (Robinson *et al.*, 1970). Exchangeable sodium enhances swelling and cracking so that water penetration is better. These properties lead to the important effects of water and aeration, which affect nutrient supply.

The uptake and use of urea-N by cotton, wheat and sorghum in the Gezira soil depend to a large extent on the depth and time of fertilizer application. High yields of these crops require the maintenance of an adequate supply of nitrogen at relatively early stages of growth. Ammonia volatilization from surface-applied urea is an important mechanism of nitrogen loss, which may reduce nitrogen fertilizer efficiency in these calcareous alkaline cracking soils. Increased ammonia losses were associated with: high urease activity, high temperature, and drying and wetting cycles soon after hydrolysis of urea. Summer crops recovered less nitrogen than winter crops due to increased water loss during high temperatures and the more frequent alternate wetting and drying of the surface soil with deep cracks resulted in increased NH_4 loss. In comparison with cotton and sorghum, wheat being a winter crop, short-seasoned and quick growing, was able to absorb as much fertilizer N as possible before appreciable loss could occur.

Conclusions and Recommendations

The irrigated sector

1. A wealth of information and experience is available in this regard. Such information could be very useful to the Africa Vertisols network of the International Board for Soil Research and Management. Some leading Sudanese scientists are available to help in this aspect.
2. The state of phosphorus in the Sudan Vertisols is not well known. Further investigations are needed on the nature of the phosphates and their kinetic behaviour.
3. Most of the irrigated schemes in the Sudan are currently not functioning according to their full potential due to inadequate inputs and inappropriate management.

The rainfed sector

This sector is lacking behind as far as agricultural technology is concerned. Research data are meagre and the large-scale farmers are not conscious of sustainable farming systems. The following are some ideas that can help in the sustainability of mechanized rainfed agriculture:

1. Rough ploughing of land immediately after harvest when some moisture is still in the soil would enhance self-mulching during soil cracking and conserve moisture.
2. Seedbed preparation and sowing should be completed immediately after rains have moistened the soil appropriately. The number of days that machinery can be used in the field decreases with increasing rains, thus decreasing from 15–18 days in June to 3–5 days in August.
3. Rainfall distribution varies greatly within the rainfed area. In the northern part moisture deficiency is frequent, while in the southern part and in the relatively low areas waterlogging might become a frequent problem. Therefore, appropriate soil moisture conservation in the former and ways to dispose of excess water in the latter should be worked out. Growing of crops and varieties of short maturity periods, and drought and waterlogging tolerance are recommended.
4. Implements and tillage operations that are least conducive to soil compaction and hard panning, effective against weeds, moisture conserving and more cost-effective should be adopted.
5. The rainfed areas should be delineated agroclimatically and the research strategy should be based on this. This would require good knowledge of the soils, agroclimates and crop requirements. Very few soil series have been identified and described well.

References

Atta El Moula, M.E. (1985) *On the Problem of Resource Management in the Sudan.* Environmental Monograph Series, Institute of Environmental Studies, University of Khartoum, Khartoum, Sudan.

Ayoub, A.T. (1986) ^{15}N-labelled urea recovery by different crops in the Sudan Gezira soils. *Fertilizer Research* 9, 213–221.

Ayoub, A.T. (1998) Extent, severity and causative factors of land degradation in the Sudan. *Journal of Arid Environment* 138, 397–409.

Ayoub, A.T. (1999) Land degradation, rainfall variability and food production in the Sahelian zone of the Sudan. *Land Degradation and Development* 10(5), 489–500.

Beshir, Mahdi (1999) *Future Sudan Development: the Ecology of the War in the South and the Rehabilitation of the Semi-arid North.* A discussion paper. Department of Political Science, University of Khartoum & Friedrich Ebert Foundation, Khartoum, Sudan.

Burford, J.R. (1987) Strategies for maintenance of soil fertility. In: Latham, M. and Ahn, P. (eds) *Management of Vertisols under Semi-arid Conditions.* IBSRAM Proceedings No. 6, IBSRAM, Bangkok, Thailand, pp. 51–61.

Burhan, H.O. and Mansi, M.G. (1970) Effects of N, P and K on yields of cotton in the Sudan Gezira. *Experimental Agriculture* 6, 279–286.

El Tom, Osman A. (1987) General properties and management of Vertisols in Sudan. *Soil Survey Administration Technical Bulletin*, Wad Medani, Sudan.

FAO (1995) *Digital Soil Map of the World*. Land and Water Development Division (FAO), Rome, Italy.

FAO (1997) FAOSTAT on-line statistical service. http://apps.fao.org FAO, Rome.

Farbrother, H.G. (1987) Supplementary irrigation. In: Latham, M. and Ahn, P. (eds) *Management of Vertisols under Semi-arid Conditions*. IBSRAM Proceedings No. 6, IBSRAM, Bangkok, Thailand, pp. 267–282.

Fink, A. (1962) Kalium status of some Sudan clay soils. *Plant and Soil* 16, 293–309.

Jewitt, T.N. (1955) *Gezira Soil*. Bulletin No. 12, Ministry of Agriculture, Sudan Government, Khartoum, 83 pp.

Jones, T.A. (1957) Nitrogen studies of the irrigated soils of the Sudan Gezira. *Journal of Soil Science* 8, 211–224.

Jones, T.A. (1958) Carbon studies on the irrigated soils of the Sudan Gezira. *Journal of Soil Science* 9, 272–276.

Mahmoud, M.A. (1988) Rain-fed agriculture and cropping systems on Vertisols in Sudan. In: Jutsi, S.C., Hague, I., McIntine, J. and Stares, J.E.S. (eds) *Management of Vertisols in Sub-Saharan Africa*. International Livestock Centre for Africa, Addis Ababa, Ethiopia, pp. 27–35.

Ministry of Environment and Tourism (1996) *Towards a National Plan for Environmental Action in the Sudan* [in Arabic]. Khartoum, Sudan.

Musa, M.M. (1968) Nitrogenous fertilizer transformations in the Sudan Gezira soil. I. Ammonia volatilization losses following surface applications of urea and ammonium sulfate. *Plant and Soil* 28, 413–421.

Musa, M.M. (1970) Some aspects of nitrogen transformations in Gezira soil. In: Siddig, M.A. and Hughes, L.C. (eds) *Cotton Growing in the Gezira Environment*. Agricultural Research Corporation, Wad Medani, Republic of the Sudan, pp. 310–317.

Robinson, G.H., Magar, W.Y. and Rai, K.D. (1970) Soil properties in relation to cotton growth. In: Siddig, M.A. and Hughes, L.C. (eds) *Cotton Growing in the Gezira Environment*, Agricultural Research Corporation, Wad Medani, Republic of the Sudan, pp. 318–325.

Said, M.B. (1973) Ammonium fixation in the Sudan Gezira soils. *Plant and Soil* 38, 9–16.

Snow, O.W. (1948) Fertilizers and manures. In: Tothill, J.D. (ed.) *Agriculture in the Sudan*. Oxford University Press, London, pp. 688–698.

Tothill, J.D. (ed.) (1954) *Agriculture in the Sudan*. Oxford University Press, London.

UNEP/ISRIC (1990) *World Map of the Status of Human-induced Soil Degradation (GLASOD)*. ISRIC, Wageningen, The Netherlands.

Walsh, R.P.D., Hulme, M. and Campbell, M. (1988) Rainfall decline and its impact on hydrology and water supply in the semi-arid zone of the Sudan. *Geographical Journal* 154, 181–198.

Willcocks, T.J. (1987) Avenues for the improvement of cultural practices on Vertisols. In: Latham, M. and Ahn, P. (eds) *Management of Vertisols under Semi-arid Conditions*. IBSRAM Proceedings No. 6, IBSRAM, Bangkok, Thailand, pp. 207–220.

Vertisols Management in Tanzania

8

F.B.R. RWEHUMBIZA,[1] N. HATIBU[2] AND H.F. MAHOO[2]

[1]Department of Soil Science, Faculty of Agriculture, Sokoine University of Agriculture, Morogoro, Tanzania; [2]Department of Agricultural Engineering and Land Use Planning, Faculty of Agriculture, Sokoine University of Agriculture, Morogoro, Tanzania

Key words: East Africa, Tanzania, Vertisols, land use, constraints, soil management

Introduction

Mainland Tanzania is located between latitudes 1.5° south and 12° south and longitudes 30° east and 40° east and covers a land area of 881,300 km^2 with an additional 61,500 km^2 under inland lakes. The islands of Zanzibar and Pemba occupy another 2000 km^2. Most of the country, except along its 800 km coastline, is situated in the Central African Plateau with altitudes ranging between 1000 and 1500 m above sea level. The country borders Kenya and Uganda to the north, the Indian Ocean to the east, Mozambique, Malawi and Zambia to the south and Rwanda, Burundi and Democratic Republic of Congo to the west.

Detailed surveys of soils of Tanzania have been carried out in only a few areas. Fewer than six out of 91 districts in the country have had some of their soils surveyed. Isolated surveys covering specific areas have been carried out by irrigation and rural development projects and environmental conservation schemes.

Serious attempts to prepare a soil map of Tanzania were made by D'Hoore (1964), Hathout (1983) and De Pauw (1984). The maps produced are not detailed enough, given the very small scales used (1 : 2,000,000–1 : 4,000,000). According to De Pauw (1984), available soil reports and maps (in 1984) covered 50% of the country at various levels of detail but mainly reconnaissance. Since then, no major surveys have been carried out. With such a background, the exact extent and mapping of different soils is sketchy and approximate in most areas.

Estimates based on the Soil Map of World (FAO–UNESCO, 1977) put Vertisols as the fourth most dominant soils, covering about 5.3% of the land area of Tanzania. Ferric Acrisols, Chromic Acrisols and Ferric Luvisols account for 33, 14 and 10%, respectively. However, later studies by De Pauw (1984) indicated that Vertisols occupy 6.5% of the land area of Tanzania. Locally, Vertisols are called *mbuga*, which means 'plains'. Indigenous knowledge surveys have found that land is called *mbuga* if it has the following characteristics:

- It lies at the bottom of a catena.
- It is relatively darker and more fertile than surrounding soils.
- It is flooded during the rains and is also sticky and cracking.

Therefore, not all soils called *mbuga* in Tanzania are Vertisols. The tendency is to refer to all soils occurring in the plains as *mbuga*, which is misleading. The Vertisols can occur in the lower parts of undulating terrain or as fans at the end of flooded areas.

Only a limited number of studies have been carried out on Vertisols occurring in the western plateau (Mwanza, Shinyanga, Tabora and Mara regions) (Fig. 8.1) to investigate chemical, physical and agronomic properties (Scaife, 1968; Mowo, 1987, 1989; Kajiru, 1995). Major findings from these studies were that:

- The Vertisols of western Tanzania differ in parent materials.
- Vertisols are low in fertility and especially N.
- In-depth studies of different Vertisols are necessary so as to come up with sound management packages.

The information presented in this chapter was obtained from overlays involving different maps. Maps used included those showing soils, crops, agroecological zones, national parks, livestock distribution, general land use, vegetation, tsetse infestation, relief and drainage. An atlas by the Ministry of Education and SIDA (1987) provided a wealth of information. In the map (Fig. 8.1) showing distribution of the Vertisols, the letters 'C', 'E', 'N' and 'P' are respectively for Vertisols occurring in the coastal zone, the eastern plateau, the northern rift zone and the central plateau.

Distribution

Because of the limited information on soils and the small scale of existing maps, only areas where Vertisols occur extensively are shown (Fig. 8.1). Furthermore, it is virtually impossible to map individual Vertisol units because soil variability affects areas that are much smaller than could ever be mapped at these small scales. The scheme used by De Pauw (1984) of indicating

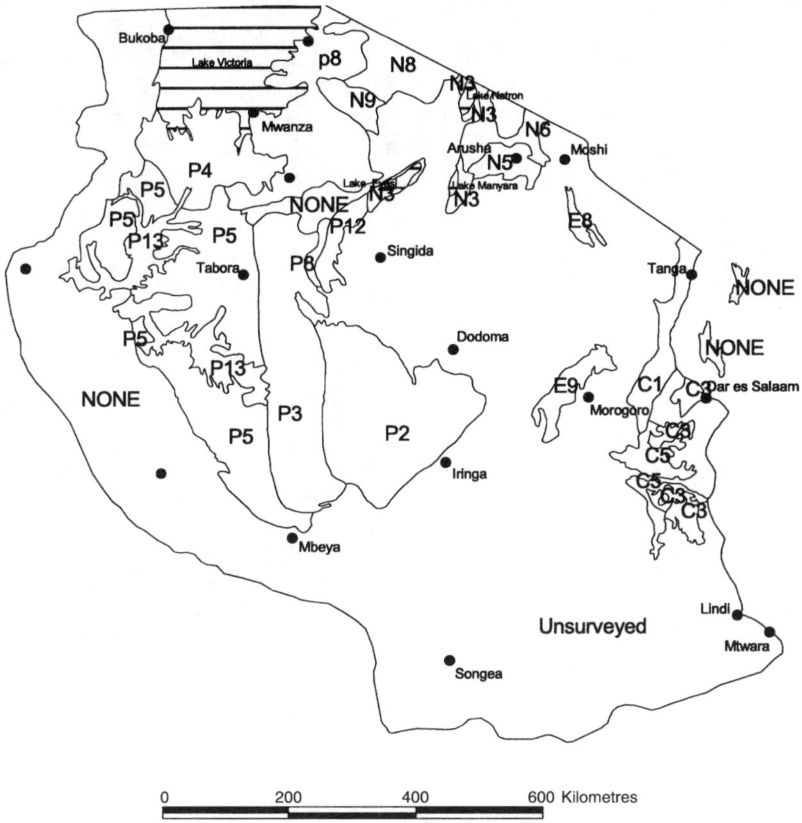

Grid:	UTM
Projection:	Tranverse Mercator
Spheroid:	Clarke 1880 (modified)
Zone:	36
Units of measurement:	Metre
Central meridian:	33 00 00 E of Greenwich
Scale factor:	0.9996
False coords of origin:	500,000m Easting 10,000,000m Northing
Datum:	New (1960 Arc)

Fig. 8.1. General distribution of Vertisols in Tanzania in relation to physiography. (Data source: E. De Pauw (1983), B. Lundgren (1975). Compiled by: F.B.R. Rwehumbiza, 1999. Cartography: RS & GIS Lab. SUA 1999.)

the proportion of each soil type in a general map unit was adopted in the presentation of Fig. 8.1 and Table 8.1.

The largest area of occurrence is the central plateau. The letter P precedes map units falling under this category. Mwanza, Mara, Shinyanga, Tabora, Kigoma and Rukwa regions have large areas covered by Vertisols. In this zone, Vertisols cover about 42,684 km^2, which is 17% of the land covered by this zone. The second largest area of occurrence is the coastal zone. Map units in this zone are preceded by the letter C. Vertisols cover about 7030 km^2 or 30%

Table 8.1. Distribution, current use and limitations to greater use of Vertisols.

Map unit	Description	Location and coverage	Area under Vertisols (%)	Rainfall regime	Current use	Limitation
C1	Coastal lowlands; mainly plains with sandy and clayey soils. Have good natural fertility	Pangani and Wami river plains and scattered pockets in Pangani and Bagamoyo Districts (in Tanga and Pwani) Total area: 8783 km^2	40	Short main growing period with very short secondary growing period	Forest reserve (native forest) Game reserve (Sadan in Bagamoyo) Crops: rice, maize, sorghum, coconut Grazing (Masai pastoralists)	Tsetse fly infestation Difficult to work with hand hoe Salinity problems in some locations
C3	Coastal lowlands; mainly strongly dissected plains and plateaus	Chalinze, Ruvu area in Bagamoyo and Kibaha Districts Total area: 8310 km^2	15	One short and unreliable growing season	Ranching (Ruvu ranch) Ruvu rice farm (irrigated) Crops: sorghum, maize at subsistence level	Tsetse flies Unreliable rains
C5	Old alluvial terraces no longer flooded from rivers	Rufiji river basin (Pwani) Total area: 8569 km^2	30	One short and unreliable growing period. Rainfall ponding and runoff collection extends the growing period	Native forest Crops: paddy (small scale) Perennial crops: coconuts, pineapples	Tsetse flies and low fertility
E8	Flood plain; affected by salinity, naturally fertile	Kifaru, Langata, Pangani-Ruvu, Nyumba ya Mungu, Mkanya-Katahe plains; Western Pare Lowland in Same and Mwanga Districts Total area: 1777 km^2	30	One very short (< 2 months) growing period and onset date mainly determined by flooding regime	Maize and paddy production with rainwater harvesting Game and forest reserve: Makanya-Katahe Livestock (Masai and Pare pastoralists)	Salinity problems, unreliable duration and onset of rains Farmer–pastoralist conflicts Vermin due to proximity to game reserve

E9	Low-altitude flat alluvial plains	Mikumi, Mkata, Mkundi, Dakawa-Wami river plains Kilosa and Morogoro Districts (Morogoro) Total area: 7701 km^2	20	One medium (4½–6½ months) growing period with unreliable onset dates. Growing period length strongly influenced by rainfall ponding and runoff collection	National Park (Mikumi) Ranching (Mkata and Dakawa ranch) Dakawa rice farm (irrigated) Maize and paddy – rainfed and rainwater harvesting Forest reserve and pastoralism	Tsetse flies Poor drainage/flooding
N3	Medium-altitude lake flats with saline soils	Areas surrounding Lakes Manyara, Natron and Eyasi in Arusha Region Total area: 3686 km^2		Very short to short growing period (2 months or less)	National parks (Lake Manyara NP and Ngorongoro Crater NP)	Poor drainage, salinity, sodicity and tsetse flies
N5	Medium-altitude volcanic plateau with fertile volcanic ash soils	Arusha, Chini and the plains surroundings Kilimanjaro International Airport (in Arusha and Kilimanjaro Regions) Total area: 4544 km^2	23	Short to medium (2–6 months) growing period with unreliable onset	Maize and beans – rainfed Limited pastoralism	Unreliable rains
N6	Medium-altitude volcanic ash soils	Monduli area Rain shadow areas of Mt Kilimanjaro and Mt Meru (Arusha) Total area: 6172 km^2	17	Short or very short growing period with unreliable onset (< 2–2.5 months)	Extensive grazing on limited pasture, no tsetse	Moisture deficit (semi-arid)
N8	High-altitude volcanic dissected to hilly volcanic plateau ash soils	Serengeti and Ngorogoro area bordering Kenya Total area: 2675 km^2	21	Short growing periods with unreliable onset (3–3.5 months)	Serengeti and Ngorongoro National Parks and Game Reserves	

Table 8.1. *Continued.*

Map unit	Description	Location and coverage	Area under Vertisols (%)	Rainfall regime	Current use	Limitation
N9	Medium-altitude volcanic plateau covered by sodic volcanic ash soils	Ngorongoro, Bariadi and Meatu plains Total area: 2675 km^2	13	Short growing period with unreliable onset	National parks and game reserves Pastoralism	Severely infected by tsetse Sodicity and salinity
P2	Gently undulating plains at medium altitude (1100–1300 m) Dark cracking clays in topographical depressions with low natural fertility, seasonal flooding	Manyoni, Rungwa (Chunya) Ruaha and Mbarali area Total area: 50,093 km^2	13	Short dependable growing period (3–3.5 months)	Sorghum, millet and groundnuts Paddy – irrigated (Mbarali rice farm) National Parks (Ruaha) Game and Forest Reserve – Rungwa	> 73% of area infested by tsetse
P3	As in P2	Nzega, Igunga Tabora and Chunya districts Total area: 42,662 km^2	9	Medium-length (4–5 months) growing periods with reliable onset dates	Paddy production, sweet potatoes Early-maturing maize, sorghum, millet Forest and game reserve	Severe infestation by tsetse
P4	Flat to gently undulating plains at medium altitude (1200–1300 m) developed on granite, banded iron stones and young alluvium, moderate fertility	Sengerema, Geita, Biharamulo–Chato area near L. Victoria Total area: 27,545 km^2	20	One medium growing season lasting 3.5–5 months Onset unreliable	Forest Game Reserve – Chato and Geita Cotton, sorghum, millet, sweet potatoes Paddy, early-maturing maize Agropastoralism (draught/animal power)	Tsetse (> 62% of area)

P5	Gently undulating plains at medium altitude. Vertisols occur in topographical depressions with low natural fertility Seasonal flooding	Kibondo, Kigoma and Mpanda districts Total area: 67,855 km^2	13	One dependable growing period of medium length (5–6 months) with reliable onset date	Maize, paddy, sorghum, millet, sweet potatoes Pastoralism (Masai) – especially in Mpanda where 600,000 herds have moved into the area since 1991 National parks (Katavi) and game reserves (Ugalla)	Tsetse in most areas limits crop and livestock production
P8	Flat to gently undulating plains at medium altitude developed partly on granites, party on old colluvium or alluvium, low fertility	Greater parts of Mwanza, Shinyanga and Mara Regions (Kwimba, Magu, Bunda, Musoma, Tarime, Maswa, Meatu and Bariadi Districts) Total area: 38,496 km^2	17	One growing season lasting 3–3.5 months Unreliable onset and duration	Agropastoralism Cotton, paddy, sorghum, millet, early-maturing maize, rainwater harvesting practised	Unreliable rains lead to low biomass to support high population of livestock
P12	Flat lowland plains developed on young alluvium with seasonal waterlogging or flooding (medium altitude)	Igunga/ Iramba districts: Seyu Sekenke flood plains		One short growing period lasting 3–3.5 months Growing period depends also on duration and depth of flooding	Some paddy production	Flooding with no possibility of water control: limited crop and livestock production (no tsetse flies)
P13	Flat seasonally inundated lowland plains. Large area under permanent or semi-permanent swamps (medium altitude)	Nikonga, Malagarasi and Ugalla river plains (Urambo and Kigoma Districts) Area: 17,216 km^2	44	As for P12 above	Limited crop and livestock production	As above, severe tsetse infestation

of the land area. In the northern rift zone and volcanic highlands represented by the letter N, Vertisols occupy 5571 km^2 or 21% of the land area. The eastern plateau (E map units) has 2073 km^2 or 22% of its land covered by Vertisols.

Vertisols thus occupy 57,358 km^2 of the land area of Tanzania. This excludes small pockets of Vertisols occurring elsewhere, which could not be shown on the map because of the small scale. Large areas of Vertisols are known also to occur in southern Tanzania (south of Ruaha/Rufiji River). No documented evidence is currently available to support this. Earlier estimates of Vertisols by FAO–UNESCO (1977) had a lower figure of 46,970 km^2 of the land area of Tanzania. Thus in the absence of detailed soil surveys to cover the whole country or even reconnaissance field checks, the exact distribution and area covered by Vertisols will remain misrepresented.

Current Use

The use of Vertisols is mainly influenced by:

- climate, and especially rainfall amount, duration and reliability;
- topography in relation to flooding and drainage;
- occurrence of tsetse flies;
- availability of farm power; and
- salinity.

Areas shown by mapping units P3, P4, P5 and P8 are the largest rice and cotton production areas in the country, especially in Mwanza, Shinyanga and Tabora regions. Additionally, the area has the largest population of livestock in the country (Fig. 8.2). Large areas of mapping units P2 and P13 are poorly drained with flooding affecting the soils for a greater part of the year. Game and forest reserves and livestock production are thus the main uses.

In the coastal zone (C1, C3 and C5), rice production, ranching, dairy farming, game reserves and pastoralism are the main activities. Large farms and ranches are found in this zone. Pastoralists have also moved large numbers of livestock into the area, mainly to take advantage of assured markets in the coastal cities. The eastern zone (E) has uses almost similar to those found in the coastal zone.

National parks (Lake Manyara, Ngorongoro Conservation Area and Serengeti) and Serengeti game and forest reserves are found on the northern plateau, specifically in N3, N8 and N9. Pastoralism is the major activity in N5 and N6 map units. The area of national parks and game reserves in Tanzania covers about 12% of the total land area, most of which falls under Vertisols (Lundgren, 1975).

The most important farming systems found in the areas with Vertisols are what has been termed 'livestock-based systems' (World Bank, 1994). All have

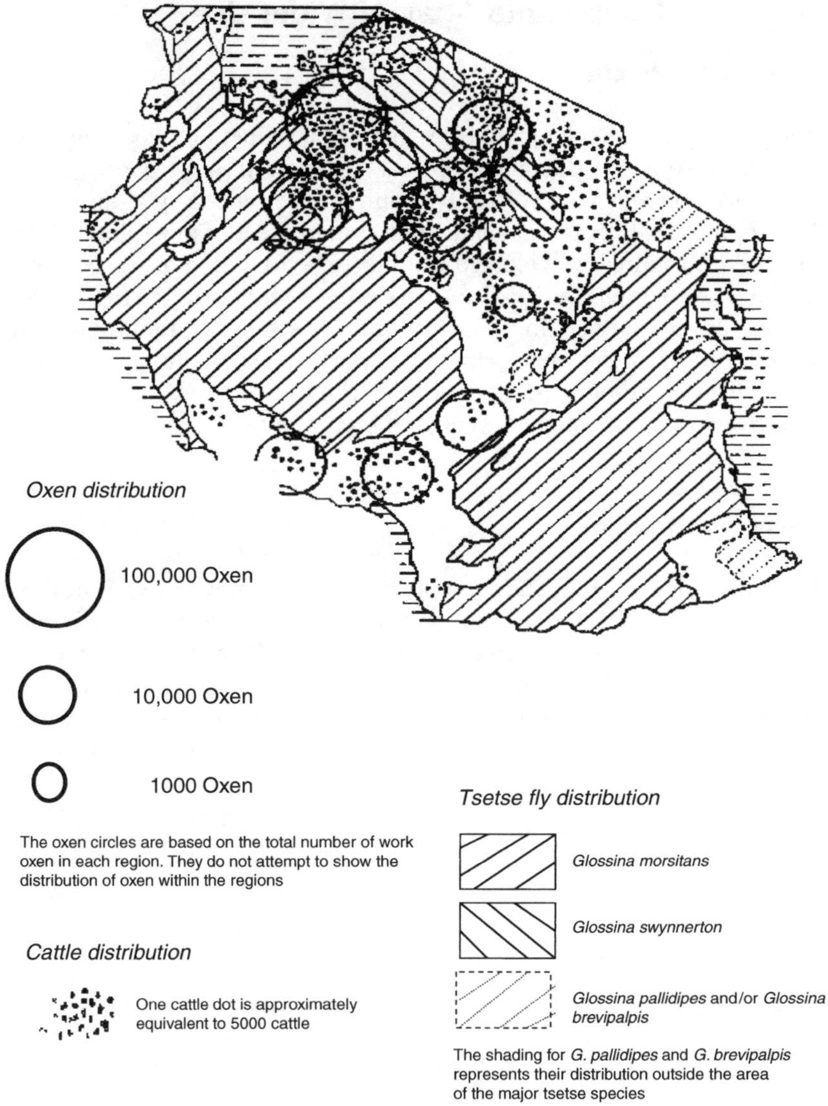

Fig. 8.2. Map of Tanzania showing distribution of work oxen, cattle and tsetse flies. (Sources: MoA (Ministry of Agriculture), 1973 (cattle and tsetse) and MoA, 1984 (oxen numbers).)

a livestock component with either: sorghum and millet, or sorghum and maize/cotton, or maize and cotton/rice or rice and maize/tobacco.

Areas with Vertisols also have the highest population of livestock in the country. For example in 1978, Mwanza, Shinyanga and Mara regions had 3.7 million cattle, 1.3 million goats and 1 million sheep (Kyomo and Chagula, 1983).

Constraints Against Wider Use

Flooding and drought

In some areas, prolonged flooding rules out crop production, especially when floods even overtop rice, the only crop that can do well under poor drainage. Examples of such areas are C3, P12, P13 and southern parts of P3 and P5.

Most of the Vertisols in Tanzania are found in areas of low rainfall (Table 8.1). However, as already explained the Vertisols are found at the bottom of the catena where runoff water collects naturally. Therefore, often the Vertisols suffer from alternating flooding and drought. Consequently, most of the time the soil is either too wet or too dry to be worked, especially using manual power. Control of runoff is therefore one of the most important factors that determine the extent to which the Vertisols can be used.

Shortage of farm power

Vertisols are difficult soils. They are sticky and heavy when wet and very hard when dry and thus require considerable draught power. The number of days when the soils can be worked is very limited. Therefore, studies have shown that shortage of farm power is the leading constraint against wider use of Vertisols for crop production (Mashaka *et al.*, 1992). In the most important farming system of livestock–maize–cotton/rice the following strategies are used:

- sowing into pits hoed on untilled soil (zero-tillage), or
- ploughing (with oxen or tractor) followed by sowing into pits made by hand hoe, or
- broadcasting seed followed by tractor or oxen-plough, or
- ploughing (with oxen or tractor) followed by broadcasting.

In general, zero tillage is the most common practice for maize and cotton (Hatibu and Mtenga, 1996). The actual loosening of soil takes place after crop establishment, during the first weeding operation. In rice production, the establishment of paddy fields is a time-consuming and labour-demanding operation. This operation is done over a period of three or more seasons. The wide adoption of draught animal power, which has been recorded in the districts dominated by Vertisols, is a clear indication of the central role of farm power in the use of these soils.

Salinity

Salinity problems occur mainly in the arid and semi-arid zones. In these environments, precipitation is usually insufficient to meet the evapotranspiration

(ET) demand. Vertisols are usually found in lowlands of alluvial plains and valleys. These flattish or internal drainage areas receive a lot of runoff from surrounding elevated areas during the rainy season. Salts accumulate as a result of continuous collection of drainage water. Since the drainage is poor, water evaporates and salts are left behind. Map units E, N and P are in semi-arid areas and are thus affected by salinity. In the coastal zone, salinity is caused by the ingress of seawater through tidal waves or underground aquifers.

Soil fertility

Vertisols are generally low in plant nutrients. Samki (1985) reported low levels of N in some Vertisols in Tanzania. Some Vertisols in Mwanza region have as low as 0.06% total N (Mowo, 1989). Current crop yields on Vertisols are very low, due to non-use of fertilizers and especially N. Factors contributing to low use of fertilizer include high prices following the removal of subsidies in 1995. Absence of an input distribution network means that fertilizers are not available when needed. Unreliable rainfall also means that the use of fertilizers can cause a salt effect on crop physiology and hence water stress in dry spells.

Tsetse infestation

Grazing has been the most important land use for Vertisols until the very recent development of rice cultivation. The widespread occurrence of tsetse makes it impractical to keep livestock in these areas. Pastoralists have resorted to burning of all vegetation as a means of controlling the problem. This has caused extensive land degradation. Therefore, despite the vast rangelands, the livestock sub-sector is performing poorly and its contribution to agricultural and national GDP is relatively low.

Management

Water for crops and livestock

Traditionally, most of the Vertisols have been cotton-growing areas. However, in recent years, there has been a rapid increase in rice production. Rice replaced cotton mainly for the following reasons:

- Growing cotton puts farmers at the mercy of the official (private and government) marketing organizations, which sometimes give very low prices or do not pay promptly.
- Cotton requires greater agricultural inputs (especially pesticides) than rice.
- With falling prices for cotton, rice generates considerably higher profits.

Rice production on Vertisols is under either rainfed or conventional irrigation water management. In the rainfed system, the rice is grown in bunded fields (locally called *jalubas*) where water is allowed to pond. Diversion ditches are used to harvest water from seasonal streams or upland areas and lead it to the bunded fields located towards the bottom of the slope. The *jalubas* are constructed by digging to a depth of about 0.2 m and the scooped soil is used to build a bund around the perimeter to a height of about 0.3 m above the ground level. Further digging and bunding continue with subsequent weeding where soil and plant material are removed and heaped on the bund to increase its size. Subsequently the bund reaches to a height of about 0.7 m (Hatibu *et al.*, 1997).

Under the conventional irrigated system, water is diverted from perennial rivers/streams to the fields through a series of canals. The major production constraints under this system include, among others, poor water control due to lack of regulatory structures, poor water distribution at the field level and poor or lack of drainage facilities.

Water supply for livestock is a problem during the dry season. During the 1950s, in an effort to improve stock supply, the government implemented projects on rainwater harvesting with storage in a number of regions of central Tanzania. These were constructed as public water supply schemes with little transfer of technology for operation and maintenance. They were therefore maintained sometimes through enforcement of strict by-laws. In the late 1960s there was laxity in the enforcement of these by-laws; this caused rapid deterioration of the storage reservoirs due to siltation. Currently, at the peak of dry season, animals are watered at pools in the streambed and reservoir beds, which are dug deeper as the dry season progresses.

Agricultural inputs and farm power

Generally fertilizer use is very limited. Additionally very few fertilizer trials have been conducted on Vertisols. A review of literature by Mowo (1989) on response to fertilizer use on Vertisols indicated that maize and cotton respond to N but not P application in some Vertisols.

Kajiru (1995) reported significant increases in rice yields from an average of 3.1 to 3.9 t ha^{-1} by applying urea at 30 kg N ha^{-1} in Kwimba (Mwanza) and Maswa (Shinyanga) Districts. However, at the same time surveys by Meertens *et al.* (1991, 1992) reported rice yields as high as 4 t ha^{-1} in parts of Kwimba District with no fertilizer input.

Farmers keep large numbers of livestock, which produce adequate manure. It has been shown through trials that application of 10 t ha^{-1} of manure would be beneficial in increasing crop yields (Hatibu *et al.*, 1997). In some areas, farmers already have adequate transport (ox-carts), which is used to transport nearly 70% of the crop from the field. Means for transporting manure are thus available, but farmers rarely use it. It might be that the farmers are not yet convinced that the maintenance of soil fertility is necessary or maybe labour rather than land productivity is more important. As long as there is enough land to maintain production by extensive means, and input/output price relations are unfavourable, farmers are unlikely to give fertility management priority.

Elaborate practices for controlling crop pests and diseases are only implemented for cotton. The ultra low volume spraying pump powered by dry cells is commonly used. Recent reduced use of these pumps has been due to the high operating cost and especially the price of dry cells. The ever-increasing price of pesticides is also partly responsible for the decline in their use and that of production of cotton, in favour of paddy rice.

Draught animal power (DAP) is the major source of farm power in most of the Vertisol areas of Mara, Mwanza, Shinyanga, Arusha, Singida and Tabora (Fig. 8.2). The distribution has a very close overlap with areas that have both a large cattle population and a high proportion of land covered by Vertisols. The areas that have cattle but no draught animals are those with only little or no areas of Vertisols. In the livestock–maize–cotton/rice farming system the use of draught animal power has become part of the system. The main limitation that still exists is that farmers only use the mouldboard plough. Implements for other operations are not used at all. Therefore weeding has become a major production constraint to those farmers who have adopted DAP. However, the use of carts for transport is widespread. There is therefore a need for promoting equipment packages, especially for paddy rice production, so as to improve productivity of both land and labour.

Range management

Most of the vegetation on the Vertisols is a combination of open mixtures of trees, shrubs and grasses, which are sometimes referred to as savannahs. The most important ecological features of the savannahs are the dominance of grasses. The most widespread savannah types are characterized by broad-leaved small trees, of which the species-rich genus *Combretum* dominates, often together with *Terminalia*. Management of the grazing lands and pastures is non-existent or poor. This is because the grazing lands are mostly communal or public resources. The only management done is through clearing of bushes and uncontrolled bush fires. These practices are believed to eradicate tsetse and other pests. Bush fires are also used as a means of facilitating good growth of pasture grass at the beginning of the rainy season. Inadequate watering points are a major problem, which leads to overconcentration of livestock in a few areas, leading to land degradation through overgrazing.

Concluding Remarks

Although published studies indicate that Vertisols are low in plant nutrients, current crop yields, especially for rice, are medium to high. Seasonal runoff from higher areas probably delivers dissolved plant nutrients. As water ceases to be a limiting factor through extensive adoption of rainwater harvesting, fertilizer inputs will be crucial to maintain or increase crop yields at or beyond the present levels. Extensive soil analyses and fertilizer trials are thus necessary to promote and guide future fertilizer use.

Additionally, only sketchy and ad hoc information exists regarding the distribution, current use and management of Vertisols in Tanzania. There are no projects or activities directly focusing on the management and efficient use of the Vertisols. However, Vertisols are important soils, which are largely used for national parks and game reserves as well as for the production of rice and cotton. Substantial information on Vertisols exists as secondary information in the literature concerning cotton, rice, national parks, woodland and game reserves.

Therefore, collection and review of this literature will generate a wealth of information about Vertisols in Tanzania. This will also create a base on which more focused studies on Vertisols can be designed.

References

De Pauw, E. (1983) *Soils and Physiography, a Map of Tanzania*. A crop monitoring and early warning systems project, GCPS/URT/047/NET. Ministry of Agriculture–FAO, Dar es Salaam.

De Pauw, E. (1984) *A Report on Soils, Physiographic and Agro-ecological Zones of Tanzania*. A crop monitoring and early warning systems project, GCPS/URT/047/NET. Ministry of Agriculture–FAO, Dar es Salaam.

D'Hoore, J.L.E. (1964) *Soils Map of Africa*. Commission for Technical Cooperation in Africa, Interafrican Pedological Service, Joint Project No. 11, Lagos.

FAO–UNESCO (1977) *Soil Map of the World*, Vol. 6, Africa. Unesco, Paris, 299 pp.

Hathout, S.A. (1983) *Soils Atlas of Tanzania*. Tanzania Publishing House, Dar es Salaam, 49 pp.

Hatibu, N. and Mtenga, N.A. (1996) *AGROTEC (Agricultural Operations Technology). Participatory Baseline Study: Smallholder Technological Constraints in Shinyanga Region, Tanzania. The Case of Bulambila Village*. FAO, Rome, 101 pp.

Hatibu, N., Lazaro, E. and Mahoo, H.F. (1997) *Farming Systems Assessment of Rain Water Harvesting for Crop Production in Tanzania*. Soil-Water Management Research Group, Sokoine University of Agriculture, Morogoro, Tanzania, 82 pp.

Kajiru, G.J. (1995) Study on the non-acceptance of fertilizer recommendations in Malya rice valley, Lake Zones, Northwest Tanzania. MSc Crop Science dissertation, Wageningen Agricultural University, The Netherlands, 52 pp.

Kyomo, M.L. and Chagula, A. (1983) Role of livestock on organic farming. In: *Proceedings of the Workshop on Resources Efficient Farming Methods for Tanzania*, May 16–20, 1983, Faculty of Agriculture, Forestry and Veterinary Science, University of Dar es Salaam, Morogoro, Tanzania. Rodale Press (USA), Emmaus, Pennsylvania, pp. 42–46.

Lundgren, B. (1975) *Land Use in Kenya and Tanzania*. The physical background and present situation and analysis of the needs for rational planning. Royal College of Forestry, International Rural Development Division, Stockholm, Sweden, 354 pp.

Mashaka, R.I., Anania, J.B., Busungu, D.G. and Braum, P.M. (1992) *Farming Systems Baseline Survey for Shinyanga Region*. MWA/GTZ IPM project, Shinyanga, Tanzania, 49 pp.

Meerteens, H.C.C., Ndege, L.J. and Enserick, H.J. (1991) *Results of the Rice On-farm Trial using Farming System Perspective: Maswa District 1989–90*. Field Note No. 13. Tanzania–Netherlands FSRP, Lake Zone, Mwanza, Tanzania.

Meerteens, H.C.C., Ndege, L.J. and Enserick, H.J. (1992) *Results of the Urea Demonstration On-farm Trial Using Farming System Perspective: Maswa District 1990–91*. Field Note No. 26. Tanzania–Netherlands FSRP, Lake Zone, Mwanza, Tanzania.

Ministry of Education and SIDA (1987) *Atlas for Primary Schools, Tanzania*. Macmillan Publishers, Basingstoke, 65 pp.

Mowo, J.G. (1987) Occurrence and management of Vertisols for the production of cotton and other crops in western Tanzania. In: *Management of Vertisols under Semi-arid Conditions*. IBSRAM Proceedings No. 6, IBSRAM, Bangkok, Thailand, pp. 138–143.

Mowo, J.G. (1989) Some properties of Vertisols of western Tanzania. In: *Vertisol Management in Africa*. IBSRAM Proceedings No. 9, IBSRAM, Bankok, Thailand, pp. 159–164.

Samki, J.K. (1985) Vertisols: their properties, classification and management in Tanzania. In: *Proceedings of the Fifth Meeting of the Eastern African Subcommittee for Soil Correlation and Land Evaluation* (Sudan, 1983). World Soils Resources Report No. 56. FAO, Rome, pp. 70–83.

Scaife, M.A. (1968) Maize fertilizer experiment in Western Tanzania. *Journal of Agricultural Science* 70, 209–222.

World Bank (1994) *Tanzania – Agriculture – a World Bank Country Study*. World Bank, Washington, DC, 250 pp.

9 Vertisols Management in Zambia

N. MUKANDA[1] AND A. MAPIKI[2]

[1]*Mt Makulu Central Research Station, Chilanga, Zambia;*
[2]*Department of Agricultural Engineering and Land Use Planning, Faculty of Agriculture, Sokoine University of Agriculture, Morogoro, Tanzania*

Key words: Southern Africa, Zambia, Vertisols, soil management, soil physical properties, soil chemical properties, yields

Introduction

Vertisols, according to the FAO–UNESCO legend (FAO, 1990), are soils having, after the upper 18 cm have been mixed, 30% or more clay in all horizons to a depth of at least 50 cm; they develop cracks from the soil surface downward, which at some period in most years (unless the soil is irrigated) are at least 1 cm wide to a depth of 50 cm. Vertisols have intersecting slickensides or wedge-shaped or parallelepiped structural aggregates at some depth between 25 and 100 cm from the surface, with or without gilgai.

According to Magai (1985), Zambia has all the ten soil orders, although to a varying extent. Using the 1983 soil map of Zambia (scale 1 : 2,500,000) to estimate the occurrence of soil orders and soil families of *Soil Taxonomy*, the extent of soil orders in Zambia, according to Magai is summarized in Table 9.1; the total area of Vertisols is estimated to be 5 million ha, representing about 7% of the total land mass and 10% of the arable land.

Vertisols are heavy, strongly cracking, clay-textured, usually dark-coloured, and often containing lime concretions at some depth in the subsoil. They occur in the lower river valleys of the Luangwa, Zambezi and Lukusashi rivers, and also in the Kafue Flats. Vertisols are most common in seasonally waterlogged depressions in the medium- and low-rainfall zones of Zambia (< 1000 mm) with altitudes being less than 1000 m above sea level.

Vertisols develop from a wide range of parent materials but are commonly derived from calcareous rocks. Vertisols are also typically developed on alluvial material in flat land areas or depressions. Typic Pellusterts and Typic

(eds J.K. Syers, F.W.T. Penning de Vries and P. Nyamudeza)

Table 9.1. Extent of soil orders in Zambia (Magai, 1985).

Soil order	Extent (%)	Soil order	Extent (%)
Entisol	25	Vertisol	7
Inceptisol	24	Spodosol	3
Oxisol	15	Mollisol	< 1
Ultisol	15	Aridisol	< 1
Alfisol	10	Histosol	< 1

Chromusterts, both being fine, mixed, isohyperthermic, are the most commonly occuring families.

The known locations of Vertisols are the Kafue Flats, Mambova swamps and Luangwa and Gwembe valleys. The Kafue Flats are a wide flat flood plain of the Kafue River in the centre of which an area is underlain by limestone and schist. Within this area soils in the depressions are usually Vertisols or Vertic Cambisols. These areas are flat with slopes of less than 1%. Gilgai micro-relief is usually common with vertical displacements varying from 10 to 25 cm depending on the expanding lattice clays contained in the soil.

Mambova swamps are undeveloped and hardly studied with solodized Vertisols formed in the Zambezi alluvium. The Luangwa and Gwembe valleys are formed by the Luangwa and Zambezi rivers. In these areas Vertisols have formed in more recent times.

Degraded Vertisols (Brammer, 1973) have been found. These soils have lighter textured topsoils and pH values of 5.0–6.0. They are found in areas bordering the Kafue Flats (e.g. Lochinvar) and the Zambezi flood plain near Mongu in Western Province.

The vegetation associated with Vertisols varies. Some of the common species are *Colophospermum mopane* woodland, *Setaria ciliolata* grassland, *Acacia gerrardii* woodland, *Oryza barthii* grassland, other acacia species or open grass. Termite mounds are common to few, the distances between them being in the range of 20–150 m. The termitaria may be as high as 1.5 m with bases of 3.5–6 m.

Methods

Soil characterization

Zambian Vertisols are the least studied and therefore no attempt here will be made to give a detailed classification. All Vertisols which have been described are situated in the Kafue river basin (FAO, 1968; Brammer, 1973; Mukanda *et al.*, 1996), an area of the greatest agricultural potential because of its large size and high concentration of people.

Management of Vertisols

The first work on the development of Vertisols in Zambia was started in 1956 on a small scale with the help of the Dutch engineering consultants Nedeco (Kerkhoven and Jelley, 1963). This involved the construction of polders at Nanga in Mazabuka to control flooding as it is done in The Netherlands. Initially an area of about 300 ha was chosen within the Kafue Flats and an earth dyke was constructed around it to control flooding.

Kerkhoven and Jelley (1963) in an effort to discover effective and efficient methods of irrigation observed that sprinkler irrigation was expensive, contrary to their original belief, and developed a good method locally known as the 'furrow and crack' method of surface irrigation. In this method water was allowed to run laterally, entering and filling the cracks down to the uncracked and almost impermeable layer, which lessens the further loss by deep percolation. Irrigation by furrow was spaced at 2–6 m apart. To increase water availability to young plants, the furrows were blocked at intervals to flood the soil surface.

Another technique developed is to construct steep and narrow permanent cambered beds inside levelled basins. The basins are then flooded with water and excess water is diverted for other uses. The steep cambered beds allow sufficient surface and internal drainage.

Results and Discussion

Soil characterization

Not all the properties of the Vertisols characterized have been studied. In fact some information on the properties that are important in irrigation management (e.g. water infiltration rates and hydraulic conductivity) was lacking until the detailed characterization of Nanga farms by Mukanda and colleagues in 1996.

Although the soils investigated are from one area, the properties discussed below are believed to be common to other Vertisols in other parts of the country as all the soils occur in similar environments.

Physical properties of Vertisols

The Vertisols in Zambia generally have very dark-coloured topsoil. The colours range from very dark grey or black in the topsoil to very dark greyish brown, greyish brown, dark grey and grey in the subsoil. The clay content is between 30 and 80%, and increases with depth at most locations (Table 9.2). The texture varies from sandy clay, sandy clay loam, clay loam in the topsoil to clayey in the subsoil.

The soil profiles studied had extremely well-developed structures, more so in the subsoil. The structure was usually weak to moderate fine, to medium sub-angular to angular-blocky, to granular in the surface layers. Below

Table 9.2. Some physical characteristics of a fine, mixed Chromustert. (Source: Woode, 1985.)

		Texture			Bulk density		Water content	
Depth (cm)	Horizon	Clay	Silt	Sand	1/3 bar	Oven dry	1/3 bar	15 bar
0–8	A	50.6	19.4	30.0	1.40	1.73	26.5	17.4
8–24	BW	52.6	21.0	26.1	1.55	1.93	23.1	16.2
24–74	BK1	52.4	20.7	26.9	1.57	2.10	23.3	17.2
74–105	BK2	53.8	20.9	25.3	1.57	2.03	24.3	19.1
105–133	BK2	56.0	19.0	25.0	1.53	1.99	26.1	19.3
133–154	BCK	56.0	18.6	25.4	1.53	1.99	27.0	20.0

the surface soil, the structure was medium to coarse sub-angular blocky to moderate coarse prismatic structure, which became stronger in the underlying layers. Slickensides were common to many. Deeper still the soil became massive or structureless. All the soils were hard or very hard when dry and friable when moist. In their wet conditions, the soils were plastic and sticky. These soils shrink and swell with changes in moisture content.

The Vertisols investigated had a high water content due to their high clay content. Most of this water, however, was not available to plants. Because both the field capacity and permanent wilting point were high, available water content (AWC) was generally low in contrast to other soils with high clay content. The AWC was between 120 and 160 mm m^{-1} soil depth. The water infiltration rates were very high (58.0 cm h^{-1}) when the soil was dry due to the presence of deep cracks. Once the soil was wet and saturated the infiltration rates became extremely low. Infiltration rates as low as 0.05 cm h^{-1} were not uncommon.

Chemical properties of Vertisols

According to Bunyolo *et al.* (1985) the range in pH conditions was mainly caused by the parent material. Most Vertisols characterized were derived from basic rocks such as basalts and limestone. Such soils were usually slightly acid to neutral. Very few were derived from acidic rocks and consequently the pH was acidic. Calcium (Ca) concretions were common in most profiles at greater depth and when they occurred the pH was usually moderately alkaline (Table 9.3). A gypsiferous horizon was present in some profiles.

The cation exchange capacity values varied considerably within the Vertisols but ranged between 30 and 80 cmol(+) kg^{-1} clay (Table 9.3). Although the higher values were most common, some more kaolinitic Vertisols were found at Nakambala Sugar Estates near Mazabuka. Generally, however, montimorillonitic clay with values above 30 cmol(+) kg^{-1} clay were dominant. With very little or no exception Vertisols in Zambia were dominated by Ca on the exchange complex. Magnesium (Mg) levels were also high (Table 9.3). Sodium (Na) contents were not high enough to make the soil saline. The exchangeable Na was usually around 16%. Some of the Vertisols in the Kafue Flats, however, were saline in the deep subsoils. The base saturation varied widely, ranging from 30 to 100%.

The soils were generally low in the major nutrients, total N (Table 9.4) and P. This was certainly one of the major limitations to crop production. In order to have sustained crop production fertilizers have to be applied to supply the major plant nutrients (Table 9.5). Despite their dark colours, these soils were generally low in organic matter content (Table 9.4). The low N levels in the soil can be attributed to low organic matter. The organic matter was highest in the topsoil and lowest in the subsoil.

Management of Vertisols

Although Vertisols are little used for agricultural production, they have a high potential once management problems are overcome. Crops such as

Table 9.3. Some chemical characteristics of a fine, mixed Chromustert. (Source: Woode, 1985.)

Depth (cm)	pH-$CaCl_2$	NH_4OAc extractable bases ($cmol(+)\ kg^{-1}$) Ca	Mg	Na	K	CEC	Exchangeable Na (%)	SAR	BS (%)
0–8	6.2	18.1	11.0	0.2	0.7	30.9	1	1	97
8–24	6.8	17.7	10.0	0.6	0.2	29.3	2	2	97
24–74	7.7	19.2	12.1	2.5	0.2	28.6	8	8	100
74–105	8.0		15.7	6.0	0.2	28.7	17	13	
105–133	7.9		16.0	7.6	0.2	27.9	17	10	
133–154	7.9		14.5	8.0	0.2	28.5	16	11	

CEC, cation exchange capacity; SAR, sodium adsorption ratio; BS, base saturation.

Table 9.4. Some plant nutrients contained in a Chromustert of the Kafue Flats. (Source: Woode, 1985.)

Soil depth (cm)	Horizon	Organic C (%)	Total N (%)	Dithionite-citrate extractable Fe	Al	Mn
0–8	A	2.45	0.186	2.3	0.3	TR
8–24	BW	0.81	0.067	2.7	0.4	TR
24–74	BK1	0.57	0.045	2.6	0.3	TR
74–105	BK2	0.56		2.6	0.3	TR
105–133	BK2	0.47		2.5	0.3	TR
133–154	BCK	0.36		2.5	0.3	TR

TR, trace.

Table 9.5. Ratings of the land quality of a Chromustert of the Kafue Flats.

	Management level	
Constraint	Low inputs	High inputs
Potential constraints		
High temperature	Severe	Severe
High humidity	Slight	Slight
Frost	Slight	Slight
Length of crop growing period (70% prob.)	120 days	120 days
Effective minimum rainfall (70% prob.)	Slight (325 mm)	Slight (325 mm)
Occurrence of critical periods	Slight	Slight
Occurrence of droughts	Slight	Slight
Soil constraints		
Oxygen availability (CO_2 toxicity)	Moderate	Slight
Potassium	Moderate	None
Boron	Slight	None
Manganese	Slight	None
Zinc	Moderate	None
Sulphur	Slight	None
Physical rooting conditions	Severe	Severe
Heavy topsoil	Severe	Severe
Flooding	Slight	None
Workability	Severe	Moderate
Mechanization/land preparation	Slight	Slight
Physical degradation	Moderate	Slight

sorghum, maize, millet, sugarcane and cotton have been known to grow well on Vertisols elsewhere. In Zambia the development of Vertisols will contribute to an increase in food production.

Due to limited knowledge of managing the soils at that time, the first year of the Kafue Pilot Polder project faced a lot of problems. The planned crop rotation of 3 years irrigated grass ley and then 3 years of double cropping with rice in the rainy season and irrigated wheat or barley in winter was not successful. There were no rice yields obtained due to serious weed infestation and pest problems and only the winter cereals gave moderate yields.

The serious problems encountered in the first year of the project were weeds, pests, waterlogging and breakage of tillage implements. The major limitations to crop production during the initial stage of research were irrigation, drainage, soil structure, cultivation, fertility and weeds.

While the impermeable subsoils increase irrigation efficiency by reducing deep percolation, they present serious internal drainage problems. The soils become very wet in the rainy season and, to have good crop yields, surface drainage by ridges and furrows and open drains is necessary. All crops except rice fail to grow in waterlogged conditions. Sub-surface drainage seems to be impractical because of the slow permeability. At the Kafue Polder, ridging was not only found to be a solution to surface drainage but also helped to lower the water-table down to more than 25 cm below the soil surface. The water was allowed to run off in the furrows. Ridging was to be done after the land had been well levelled and graded.

In their natural state, Vertisols are compacted. This tends to produce big clods when the land is ploughed in the dry state. To break up the clods to a fine tilth, a lot of power is required. Ploughing should be done when the soil is at optimum moisture content. If the soils are too wet, ploughing is hampered as they puddle and stick to implements. Once the land is ploughed, the soil structure tends to improve with cultivation. Due to the non-development of these soils on a large scale very little can be said about their cultivation requirements. At Kafue Polder, the implements available were medium-powered wheeled tractors and normal ploughs.

The problems of tillage of these soils would be more serious to small-scale farmers, who normally use simple hand implements. Because the soils dry fast, farmers using simple tools would not be able to till the soil at its optimum moisture content. The massive clods would present problems in providing a fine seedbed.

Although the soils in the Kafue Flats have a high fertility status, a regular supply of the major plant nutrients (N, P and K) is needed. Kerkhoven and Jelley (1963) showed strong responses to N fertilizers by non-leguminous crops.

The problem of weeds becomes less serious once surface and internal drainage is adequate. This will allow accessibility to the fields to control weeds with herbicides.

The yields of some crops obtained at Kafue Polder are given in Table 9.6. Since then much more high-yielding varieties have been developed. The crop yields as shown are high and are comparable to yields obtained on other soils. The yields will be much higher with the development of improved management techniques.

The Kafue Pilot Polder was abandoned in the early 1960s and since then no further research has been followed up. A few farmers, however, are growing crops on Vertisols. They are all aware of the management problems.

The returns in terms of yields were not as high at Nakambala. A crop rotation to improve soil condition was established involving soybean – winter wheat – soybean – 6–7 years sugarcane.

Soybeans grew more successfully, yielding 1440 kg ha^{-1} in the 1981/82 season and increasing to 1728 kg ha^{-1} in the 1982/83 season. Cotton achieved a yield of 1376 kg ha^{-1} while that of sugarcane was 108 t ha^{-1} after 1.5–2 years. The sugarcane yield in 12 months was about 80 t ha^{-1}. On well-drained soils the yields can be as high as 233 t ha^{-1} with an average of 112 t ha^{-1}.

The Commonwealth Development Co-operation Farm with an interest in developing Vertisols goes through about 11 operations before planting. The main purpose of these operations is to achieve good levelling to eliminate

Table 9.6. Crop and experimental yields (in kg) on the Kafue Pilot Polder.

Winter crops (kg ha^{-1})		Summer crops (kg ha^{-1})		Perennial crops (kg ha^{-1} year^{-1})	
Wheat	4480	Cotton	4700	Clover (hay)	4,480
Barley	4480	Soybean	3600	Lucerne (hay)	12,300
Peas	2800	Groundnut	3600	Sugarcane	112,000
Beans	2240	Rice	7800	Grasses (DM)*	4,480
		Maize	7800		
		Kenaf	4300		
		Sunflower	4480		

*Grass species are *Paspalum*, *Chloris* and *Setaria*.

waterlogged depressions and to improve the tilth. The cost of the complete land development procedure was calculated at roughly US$400 ha^{-1}. The land has now been cropped for the past 3 years, obtaining negligible yields from rice, maize and wheat. Only cotton showed modest yields, which increased every year. In the first year of cultivation 25% of potential yield was obtained, in the second year 50% and this increased to 80% of potential yield in the area in the third year. In 3 years the farmer has developed 250 ha of Vertisols at a cost of about US$400,000.

At Nakambala Sugar Estates similarly high costs of land development are experienced, averaging US$560 ha^{-1}, attributed mainly to land planning, levelling and ripping carried out by contractors. High costs of land preparation are due to the monopolistic nature of contractors and they are expected to rise.

The major limitations to crop production are summarized in Table 9.5 and are as follows:

1. Heavy clay texture and shrinking and swelling properties. Unless the soils are tilled when the moisture content is optimum, land preparation will present many difficulties.
2. Lack of major nutrients and low organic matter. Sustained yields can only be obtained if the major nutrients are supplied in adequate amounts. Crop residues should be incorporated in the soil to increase organic matter content. High organic matter content will increase biological activity and consequently improve soil physical condition.
3. Low water infiltration rates and slow internal drainage. Once the soil physical condition is improved, the water infiltration rates are increased. Surface and subsoil drainage is needed to remove water from the fields. High water-table levels are hazardous as they may bring about salinity problems.
4. Salinity. Due to low infiltration rates and low permeability rates, salinity may become a problem with continuous cultivation. Although none of the Zambian rivers are saline, accumulation of salts on the surface could become a problem with many years of cultivation. This has already been observed at Nakambala Sugar Estates, especially in the dry season. Salt accumulations are often seen in cultivated fields.

Successful production of crops will therefore depend on whether solutions to the problems stated above are found. Management techniques should be developed to suit soil conditions.

References

Brammer, H. (1973) *Soil Profile Descriptions.* Soil Survey Report No. 11. Department of Agriculture, Lusaka, Zambia.

Bunyolo, A.M., Magai, R. and Veldkamp, W.J. (1985) *Properties, Management and Classification of Vertisols in Zambia.* World Soil Resources Reports No. 56. FAO, Rome.

FAO (1968) *Soil Survey Final Report,* Vol. II, *Multipurpose Survey of the Kafue River Basin.* FAO:35/ZAM. FAO, Rome.

FAO (1990) *FAO–UNESCO Soil Map of the World.* Revised Legend. World Soil Resources Report No. 60, FAO, Rome, 119 pp.

Kerkhoven, G.O. and Jelley, R.M. (1963) *Kafue Pilot Polder Work.* World Crops.

Magai, R. (1985) The soils of Zambia. In: Woode, P.R. (ed.) *XI International Forum on Soil Taxonomy and Agrotechnology Transfer,* 15 July–1 August, 1985. USDA Soil Conservation Service, Lincoln, Nebraska, pp. 136–140.

Mukanda, N., Chindinda, M.H. and Gondwe, P. (1996) *Semi-detailed Soil Survey of Wolverton Ranch, Farm Nos. 603 a/Rem* an *121 a Rem, Nanga Farms, Mazabuka District, Zambia.* Soil Survey Report No. 223. Department of Agriculture, Lusaka, Zambia.

Woode, P.R. (ed.) (1985) *XI International Forum on Soil Taxonomy and Agrotechnology Transfer,* 15 July–1 August, 1985. USDA Soil Conservation Service, Lincoln, Nebraska, pp. 420–427.

Vertisols Management in Zimbabwe 10

P. NYAMUDEZA,[1] J. HUSSEIN[2] AND B. MATIBIRI[3]

[1]Department of Research and Specialist Services, Chiredzi Research Station, Chiredzi, Zimbabwe; [2]Department of Soil Science and Agricultural Engineering, University of Zimbabwe, Harare, Zimbabwe; [3]Zimbabwe Sugar Association Experiment Station, Chiredzi, Zimbabwe

Key words: Southern Africa, Zimbabwe, Vertisols, soil management, soil properties, irrigation, implements

Introduction

In Zimbabwe, Vertisols constitute about 5% of the land area and are mainly derived from basalt rock. They are also derived from mudstone, greenstone and epidiorite (Kanyanda, 1989). The largest area of Vertisols is found in the southeast of the country as shown in Fig. 10.1. The basalt-derived Vertisols in the southeast are classified as Chisumbanje 3B under the Zimbabwean classification system, Pellustert group in the US taxonomy and Eutric Vertisol in the Food and Agriculture Organization legend (Nyamapfene, 1991). The soils are very productive and have been described by Thompson and Purves (1978) as inherently fertile. This chapter will describe the properties and management of Vertisols in the southeast lowveld covering Ndowoyo, Sangwe and Matibi 2 communal lands, and Chisumbanje, Hippo Valley and Mkwasine Estates.

The climate of the southeast lowveld is typical semi-arid. Generally the mean annual rainfall is about 500 mm. The annual total rainfall ranges from 250 to 1000 mm and this season (1998/99) Chisumbanje has received a record high of 1227 mm. Mean summer temperatures generally exceed 30°C with maxima around 40°C, whereas mean winter values are around 15°C with only rare frost occurrence. Low *et al.* (1984) reported an overall annual precipitation deficit of some 1400 mm.

The major land use activities are cropping and keeping livestock. The major crops are sugarcane, cotton and wheat under irrigation, and

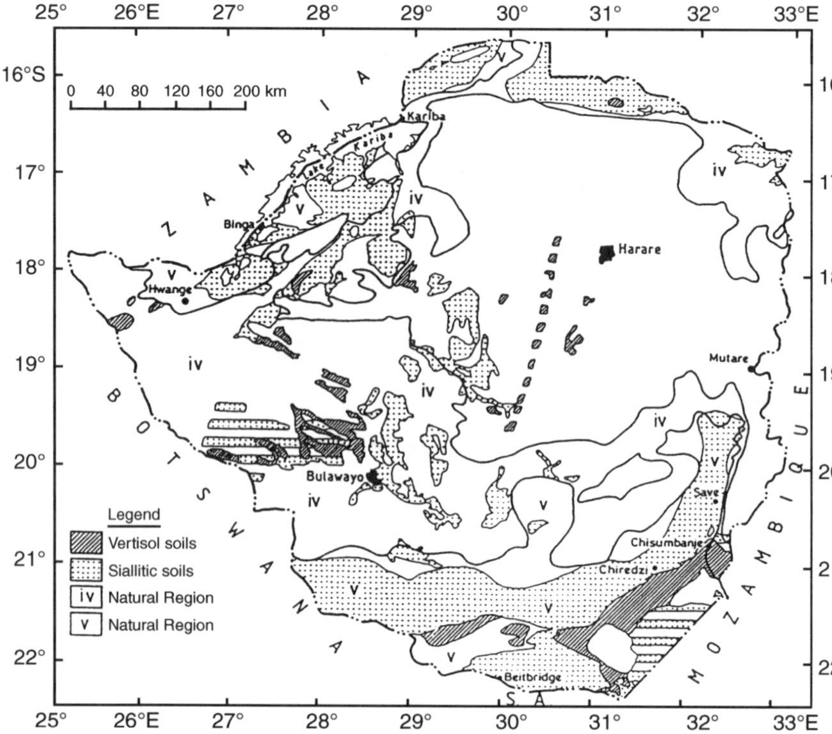

Fig. 10.1. Location of Vertisols and Natural Regions iv and v in Zimbabwe. (Surveyor General, 1979, 1980.)

maize, sorghum, cotton and sunflower under rainfed conditions. Another land-use activity that is gaining popularity is wildlife management in Gonarezhou, Malilangwe Conservancy and Mahenye Campfire Programme.

Physical and Chemical Properties

In Zimbabwe, Vertisols derived from Jurassic basalt commonly have high clay contents (> 600 g kg^{-1}); the clay fraction is composed predominantly of montmorillonite but may contain small amounts of illite, kaolinite and quartz (Nyamapfene, 1984; Ristori *et al.*, 1992). However some Vertisols with lower clay contents are found in areas where basalts overlie Karoo sandstones. The basalt-derived Vertisols are typically dark grey, brownish black or black and are classified as very fine, montmorillonitic, hyperthermic, Typic Pellusterts (Soil Survey Staff, 1975).

Vertisols can be very productive due to their strong nutrient and water-holding properties. However, management of drainage and tillage in these soils can be very difficult. The soils are very hard when dry and intractable and sticky when wet, thus tillage may only be easily performed within a limited range of moisture contents. However, the ease of tillage, infiltration, evaporation and susceptibility to erosion are very dependent on the surface structure of Vertisols (Probert *et al.*, 1987).

The surface structures of Zimbabwean Vertisols vary greatly in their morphology, ranging from fine crumb (self-mulching), typified by the Vertisols around Chisumbanje, to coarse blocky (non-mulching), as exhibited by the Jotsholo Vertisols in the northwest. Previous work (Hussein, 1981) has shown that, in coarse blocky soils, infiltration decreases rapidly with a consequent limited cumulative infiltration. However, those soils with surface crumb structure have faster and more prolonged infiltration than those soils with coarse blocky structure. Field measurements of infiltration rates under ponding conditions ranged from > 17 mm min^{-1} (in dry cracked soil) to 0 mm min^{-1} (in a wet Vertisol). The coarse blocky-structured soil shows complete surface sealing and thus limits the amount of water penetrating the soil.

This difference in infiltration is then expressed in the different water-storing abilities of the two structure types, with the crumb-structured soil holding more water at depth than the blocky-structured soil (Fig. 10.2). Plant-available water contents are, therefore, best measured *in situ* to take account of this variability, as laboratory estimates may not be reliable.

Deep drainage of Vertisols depends on the structure, texture and underlying rock. Drainage rates varying between 1 and 22 mm day^{-1} have been measured on blocky- and crumb-structured Vertisols respectively. Slow internal drainage rates are likely to cause problems in low-lying Vertisols, particularly those under irrigation, and cases of waterlogged soils and rising water-tables have been reported on a number of irrigation schemes (Hussein *et al.*, 1992).

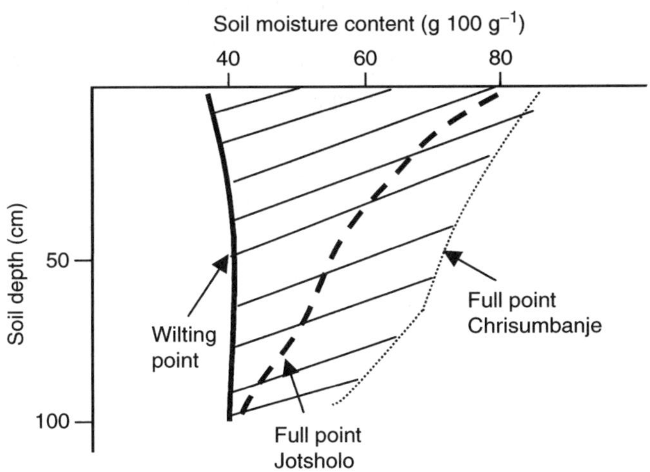

Fig. 10.2. Total available field moisture contents (TAM) in two contrasting Zimbabwean Vertisols.

Vertisols are potentially extremely erodible due to their slow infiltration rates when wet and their inherent structural instability (Mullins *et al.*, 1987). Little work on Vertisol erosion has been done in Zimbabwe to date, apart from some measurements under rainfall simulation with a nozzle producing drop size distributions and energy values of a typical high-intensity tropical storm (Hussein *et al.*, 1992). These Vertisols were even more erodible than the most degraded red clay soils, with pronounced soil loss and runoff.

In Zimbabwe, Vertisols typically have high base contents and large cation exchange capacities with alkaline pHs (Table 10.1) (Hussein *et al.*, 1992). Sodium levels are typically low but over-irrigation, coupled with poor drainage, leads to an increase in salinity and/or sodicity in these soils (Hussein *et al.*, 1992).

Organic matter levels are low, even in virgin soils, with organic carbon (C) values of less than 1.2% typically reported (Ristori *et al.*, 1992). Total nitrogen (N) levels are also usually low (< 0.1%) and most crops require N applications to produce good yields. Phosphorus levels are typically < 20 mg kg^{-1} in virgin soils, but can be increased successfully with applications of P-fertilizers (Hussein *et al.*, 1992). Levels of potassium (K) are usually adequate for many crops but under high K-consumption crops, such as sugarcane, may require maintenance applications (Nyati, 1998).

Land Use

Rainfed farming on Vertisols

As shown in Fig. 10.1 the largest area of Vertisols is located in Natural Region 5 (NR 5). The Natural Regions are a classification of the agricultural potential of the country, from NR 1, which represents the high-altitude wet areas, to NR 5, which receives low and erratic rainfall (Vincent and Thomas, 1960). According to the classification, NR 5 was said to be unsuitable for growing even the most drought-tolerant crops. However, that was not quite true

Table 10.1. Chemical properties of cultivated surface soils (0–150 mm) derived from basalt. (Source: Hussein, 1981; Hussein *et al.*, 1992.)

Property	Range of values
Exchangeable calcium (cmol(+) kg^{-1})	40–64
Exchangeable magnesium (cmol(+) kg^{-1})	16–37
Exchangeable sodium (cmol(+) kg^{-1})	0.3–7.0
Exchangeable potassium (cmol(+) kg^{-1})	0.4–2.4
Cation exchange capacity (cmol(+) kg^{-1})	64–100
pH (1 : 5 $CaCl_2$)	6.8–7.7
Mineral nitrogen before incubation (mg kg^{-1})	15–33
Exchangeable sodium per cent	< 1–7

because thousands of smallholder farmers (estimated to be about 25,000) cultivate Vertisols and grow rainfed crops in the southeast lowveld.

The three communal lands and one resettlement scheme in the southeast lowveld where rainfed farming is practised by smallholder farmers have a total area of about 450,000 ha. About 55,000 ha or 12% of the total area was cultivated during the 1998/99 season. Figure 10.3 shows the area under maize, sorghum and cotton over the past 5 years. The area under cotton has increased tremendously over the past 5 years. Another noteworthy feature in Fig. 10.3 is the relatively large area under maize and sorghum in 1995 and 1996 when free seed packs were handed out by the government.

Traditional land preparation is ploughing using draught power and if the farmer has no cattle the land is dug out by hand using hoes. Both cattle and non-cattle owners hire tractors from other farmers or the Government Tillage Unit. The crops are either dry or wet planted. Dry planting is most prevalent when rains are late. Planting starts in November and can extend up to February depending on the rains. None of the farmers apply or have ever applied organic or inorganic fertilizers to their crops. They believe that fertilizers increase crop water stress but this raises some concern about effects the practice has had on soil quality over the years.

In most years the major constraint to crop production is lack of adequate water for crop growth. The rainfall is low and poorly distributed. An analysis of the rainfall by Jones *et al.* (1987) showed that farmers in NR 5 can expect to harvest a good yield of sorghum or pearl millet one year in five, mediocre yields in three and complete failure in one. Soil fertility can also be a constraint when

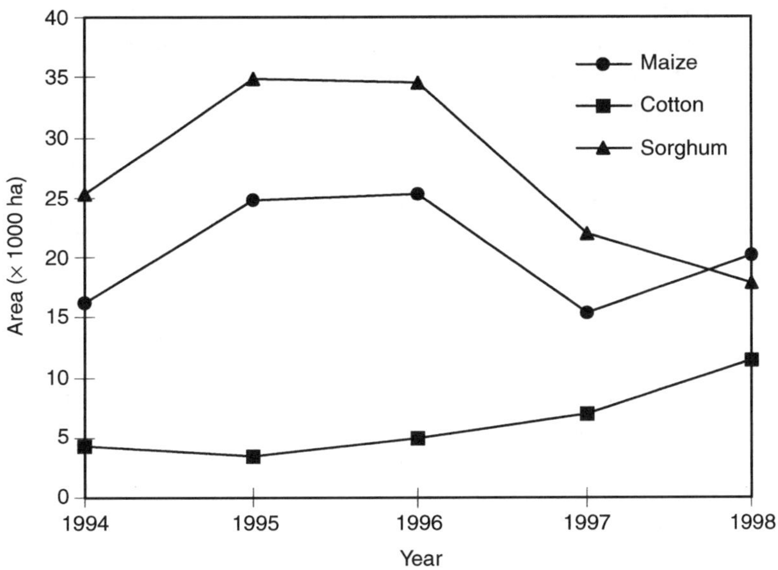

Fig. 10.3. Area under maize, cotton and sorghum in Ndowoyo, Sangwe and Matibi 2 Communal Lands over a 5-year period.

rainfall is relatively high. Nyamudeza (1998) also reported significant crop responses to nitrogen on fields that had been under cultivation for more than 30 years. However, farmers do not consider soil fertility to be a problem. A constraint appreciated by all farmers is the workability of the soil when wet. Some farmers try to avoid this problem by dry planting but weed pressure on a dry-planted crop is high. Ploughing and planting after the first flash of weeds reduce weed pressure. In good rainfall years, delays in weeding due to the sticky nature of the soil reduce crop yields.

Technological interventions and adoption in rainfed farming

Technological interventions aim at increasing and/or stabilizing crop yields. These are rainfall conservation/concentration, reducing plant populations, sowing dates and fertilizer application. There was a need to diversify crops grown in the area by introducing new crops, e.g. guar bean, and promoting crops being grown on a small scale, e.g. cotton and sunflower. The greatest effort was put into rainwater conservation using tied ridges and sowing crops in furrows, hence we called it the tied furrow technique (TFT).

Experiments on the TFT started at Chisumbanje Experiment Station in 1982. The TFT involves ploughing and ridging across the slope, preferably before the rains. The operation can be done using draught power or tractor-drawn ridgers. Since most farmers use draught power for land preparation, a joint Department of Research and Specialist Services and International Board for Soil Research and Management project (1985–1990) looked at developing and introducing ox-drawn ridgers in the Chisumbanje area. The result of the work was a modification of the wings of a ridger that was already on the market. The ridger is made of metal (all parts) and has two handles and two wings for moulding the ridges, attached to the same beam.

Table 10.2 gives a summary of the results obtained from comparing the TFT and the traditional practice of sowing crops on relatively level fields (flat). The most notable feature of the data is the variation in yield level between years, which to some extent was a direct reflection of the rainfall amounts and distribution. The response of crops to sowing in furrows was also extremely variable. The responses were small in a good year, 1984/85, and the poorest year, 1983/84. The average yield response to sowing in furrows over 6 years was 26, 27 and 45% for sorghum, cotton and maize, respectively.

In 1990, the Triangle Company was promoting cotton production in the Chisumbanje area. The company offered tractors to ridge a total of 100 ha at 1 ha per farmer in the villages surrounding Chisumbanje Experiment Station. Farmers were asked to pay a nominal fee of Zim$20. The company offered to do it free but the fee was introduced for screening purposes. About 500 farmers were prepared to pay but the first 100 to pay got the service. The enthusiasm was a clear indication that farmers had seen the benefits of the TFT. A survey carried out in 1991 (Nyamudeza *et al.*, 1992) showed that a good number of

Table 10.2. The yield of sorghum, maize and cotton (kg ha^{-1}) grown on the flat over 6 years, and the effect on yield of sowing the crop in furrows (TFT). (Source: Jones *et al.*, 1987.)

Season	Rainfall (mm)	Sorghum		Maize		Cotton	
		Flat	TFT effect	Flat	TFT effect	Flat	TFT effect
1982/83	307	730	+1626	1329	+831	1041	+1153
1983/84	360	353	+136	57	+22	82	+104
1984/85	699	3701	+57	2488	+44	2505	–67
1985/86	439	1647	+586	1821	+288	1560	+19
1986/87	275	1820	+280	515	+112	259	+18
1987/88	709	3380	+357	1590	+2194	2591	+947
Mean		1939	+507	1300	+582	1340	+362
%			+26		+45		+27

farmers had adopted the technique and 93% of the adopters claimed that by using it they had obtained an increase in crop yields. The majority of those who had adopted had been exposed to the technique through field days and experiments carried out on their fields. Unfortunately in 1990/91 and 1991/92 the rainfall was very poor and little benefit was obtained from the ridges made by the Triangle Company but some farmers who had made use of the late rains in April reported some benefits (Pushparajah, 1991). In 1991/92 Zimbabwe experienced the worst drought in history and thousands of cattle died. Thereafter, for some years, the adoption of the TFT appeared moribund, probably due to lack of draught power.

During the past 3 years a number of farmers have started to ridge and sow in furrows but do not tie or dam the furrows because of labour constraints. Farmers see the benefits as good establishment and higher crop yields. The sudden increase in the number of farmers adopting the modified TFT seems to be attributed to the sudden increase in the number of tractors in Ndowoyo Communal Lands. A tractor ownership survey conducted by Chisumbanje Experiment Station in March 1999 showed that the number of tractors increased tremendously over the past 9 years, as shown in Fig. 10.4. In 1980 there was a total of eight tractors. During the last season (1998/99), about 4000 farmers hired tractors for land preparation and 37% hired for ploughing and ridging.

Experiments on N application using moderate rates of 16 and 32 kg N ha^{-1} plus a control, zero applied N, showed little yield response to applied N. The experiments ran for three seasons using maize, sorghum and cotton. The field used for the experiments had been under cultivation for two seasons when the experiments started. Hence the poor crop response to N could have been due to the fertile virgin soil, poor rainfall or both.

In 1995/96 and 1996/97 a second set of N experiments was carried out on maize and sorghum at four sites. The N levels ranged from 0 to

160 kg N ha^{-1}. There were significant responses to N, especially in a higher rainfall year when maize yields were increased by 1.3–4.4 times (Nyamudeza, 1998). The results of the two sets of experiments clearly show that N response depends on the rainfall amount and distribution. Since the seasonal rainfall is very variable, formulation of specific fertilizer recommendations is difficult. Farmers could use the 'response farming' technique suggested by McCowan *et al.* (1992) but it may not appeal to the resource-poor farmers. In 'response farming' the timing of the onset of the rain season and the amount of early rain are used to predict seasonal rainfall and yield. To date little effort has been made on promoting fertilizer use on Vertisols under rainfed conditions. It is an area needing further investigation.

Plant population and sowing date experiments simply proved that the farmers' practice is correct. The maize population in most farmers' fields is about 15,000 to 20,000 plants ha^{-1}. The relatively low population ensures some yield in most years. Results from sowing date experiments showed that maize, sorghum and sunflowers can be sown as late as February and produce good yields if the rains end in April. However, the cool temperatures in April and May are not conducive for cotton boll formation. Farmers are aware of this and sunflower replaces cotton as a cash crop when rains are late.

Whereas before 1980 the major crops were sorghum and maize, farmers have diversified into cotton and sunflower production. After the successful performance of guar bean on station and on farm, a number of farmers grew it in 1985 but lack of a guaranteed market phased out the crop. The adoption of cotton by many farmers as shown in Fig. 10.3 is due to research work and the vigorous promotion by the private sector, especially the Triangle Company, in the late 1980s and early 1990s. At present three companies are buying cotton in the southeastern lowveld and the Cotton Company of Zimbabwe provides credit to cotton farmers.

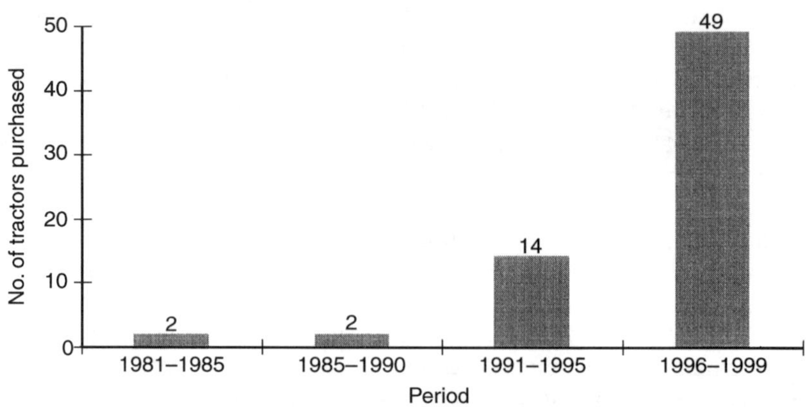

Fig. 10.4. Number of tractors purchased by farmers in Ndowoyo Communal Lands between 1981 and 1999.

Irrigated farming on Vertisols

The largest area under irrigation is in the southeastern lowveld, which has about 8200 ha. In the west of the country 760 ha are under irrigation at Jotsholo Estate. Therefore approximately 8960 ha of Vertisols are under irrigation in Zimbabwe. The major crops are sugarcane, cotton and wheat. In the 1998/99 season sugarcane covered about 6800 ha, 76% of the total irrigated Vertisol area. The total area under sugarcane in the lowveld sugar industry on both Vertisol and the other main soil type, paragneiss soil (Chromic Luvisol), was about 43,000 ha in the 1998/99 season. Of the total area of 1900 ha cropped by the 10 ha Chipiwa smallholders on Mkwasine Estate, about 300 ha are on Vertisols.

Under irrigation relatively high yields of cotton and wheat are obtained compared with other soil types. The mean cotton yield at Chisumbanje Estate over the past 10 years (1988–1998) was 1.3 t ha^{-1} with a range of 0.7–1.8 t ha^{-1}. The wheat yield is about 3.5 t ha^{-1} on average. The mean cane yield for the industry in 1998/99 was 110–115 t ha^{-1}. Mean annual cane yields on Vertisols are usually lower than the industrial average. At Mkwasine Estate for example, 500 ha on Vertisols averaged 20 t ha^{-1} lower than the 4300 ha of other soil types in 1998/99. This has been the same trend since the estate was established but plant cane yields of about 145 t ha^{-1} can be obtained.

Table 10.3 gives the area cropped and a comparison of cane yields on Vertisols and paragneiss soil. The depressed yields in the 1996/97 season were due to a partial drought, which, for example, forced Hippo Valley Estate (HVE) to limit water application to an average of 7 Ml ha^{-1} compared to 16 Ml ha^{-1} in a normal year. HVE records showed that virgin Vertisol fields out-yielded paragneiss fields for the first few years of cropping from about 1974 to 1981 before yields fell to the paragneiss level. This is probably partly related to the deterioration in Vertisol structure caused by continuous wetness due to

Table 10.3. Comparison of cane yields on Vertisols and paragneiss: Agricultural and Rural Development Authority (ARDA) in Chisumbanje, Hippo Valley Estate (HVE) and Zimbabwe Sugar Association Experiment Station (ZSAES).

	Area cropped (ha)				Cane yield (t ha^{-1})			
	Vertisols		Paragneiss		Vertisols		Paragneiss	
Year	ARDA	HVE	HVE	ZSAES	ARDA	HVE	HVE	ZSAES
1995/96	300	3000	6825	33	145[a]	101	97	135
1996/97	700	2998	6803	43	94	75	76	130
1997/98	1100	3561	8613	44	121	109	110	143
1998/99	1100	4309	7800	46	101	108	110	149

[a]Yield from first crop as opposed to ratoon crop on other estates.

irrigation as this did not allow for self-churning of these clays, which is necessary for the regeneration of soil structure and maintenance of adequate root aeration. Chisumbanje Estate was forced by prevailing circumstances to harvest a significant proportion of its cane while it was still immature at less than 12 months of age in the 1998/99 season. The Zimbabwe Sugar Association Experiment Station (ZSAES) yields, although obtained on plots which are smaller than commercial fields, give an indication of industrial yield potential from 1995 to 1999. It is interesting to compare cane yields between HVE, where cropping on most of the Vertisol area commenced over 20 years ago, and Chisumbanje, where cropping was in the fourth year in the 1998/99 season. Barring the 1996/97 drought and the immature harvest in the 1998/99 season on Chisumbanje Estate, it would appear that the trend of initial high cane yields obtained by HVE in the early Vertisol crops is the same trend unfolding on Chisumbanje. It will be interesting to follow these yield trends in the long term. Generally, sugarcane on Vertisols has given very poor ratooning yields after the fifth ratoon and replanting is necessary much earlier than on parageiss soils.

One problem identified on fields planted to cane has been the rapid decline of P and K from the virgin state to the present. The sugar industry has successfully mined K from the soil over the past 20 or more years. The change in level can be related to both what has been applied (mainly muriate of potash) if below the critical level of 0.35 K cmol(+) kg^{-1} in the soil and the yield removed, per change in K cmol(+) kg^{-1}. This can serve as a guide to the P and K replacement strategy and should be examined for individual fields on each estate. There is a need to prevent P and K from falling below threshold levels (Clowes and Breakwell, 1998). Luxury applications of P and K are wasteful and excessive K can be taken up by the cane, particularly variety N14, and this would be delivered to the mill, where increase in K would reduce the recovery of sugar.

On HVE the level of K in one Vertisol field has dropped from 1.43 to 0.66 cmol(+) kg^{-1} (0.77 units) in 17 years and produced approximately 1955 t of cane (115 t ha^{-1} $year^{-1}$). The drop has been equivalent to 0.045 K cmol(+) kg^{-1} $year^{-1}$ or per 115 t cane and the K level remains above the critical level of 0.35 below which it would be recommended that additional K should be considered.

Technological interventions and adoption in irrigated farming

In commercial land preparation, ripping is not needed as the soils are self-mulching. Normal ploughing is usually practised. Land planing is essential, especially for furrow irrigation. Harrowing is necessary but rolling is not needed, since, unlike other soils, Vertisols have suitable natural tilth.

On slopes of less than 2%, furrow irrigation is considered the most suitable but drip is also acceptable and is practised. Over-irrigation should be avoided, except when it is necessary to leach salts from the soil. Vertisols are generally less sensitive to moisture stress due to their high water-holding capacity.

Although smallholder dryland farmers do not apply fertilizers, under irrigation it is a must for optimum yields. Nitrogen applications on sugarcane, for example, range between 120 and 160 kg N ha^{-1} for a 14-month plant crop and between 140 and 180 kg N ha^{-1} for a 12-month ratoon crop. The plant crop requires less N because it is planted after a 3-month fallow period while the ratoon crop establishes soon after harvest. Fertilizer P and K applications are normally required. However, in virgin Vertisols, K may initially be present at adequate levels and in these cases fertilizer K would not be applied. Preferred phosphate fertilizer is single superphosphate (SSP) (8.3% P and 12% S), since, in addition to its primary purpose of satisfying the P requirements, SSP also supplies appreciable amounts of S. Local research has shown that cane needs about 40 kg S ha^{-1} $year^{-1}$ (Nyati, 1998). Double superphosphate (DSP) (16.2% P and 5% S), which supplies lesser amounts of S, is sometimes used.

Although Vertisols can be very productive under irrigation, problems may be experienced (Loveday, 1984) with:

1. *Slow/poor infiltration when wet.* Infiltration is good soon after ploughing but is very poor on subsequent irrigations as pore spaces have been clogged up. HVE normally does inter-row ripping after cutting to improve infiltration and also practises in-row as opposed to inter-row irrigation.
2. *Poor internal drainage, particularly on the deeper Vertisols.* Strategies are needed on how Vertisols can be drained and the depth and spacing of the drains.
3. *Poor aeration and retarded root development.* Farmers are encouraged to harvest cane on Vertisols in winter (before the end of September) when the soil is drier and the risk of compaction less.
4. *A narrow optimum moisture range for tillage, planting and harvesting operations.*
5. *Uneven land surface requiring levelling and grading.*
6. *Erosion, salinity and sodicity hazards associated with poor irrigation management.* The main irrigation systems used (furrow, drip and centre pivot) need to be evaluated for Vertisols so that the best method can be determined. However, the end-gun effect of centre pivot systems, which supply more water at the end of the pivot, seems to have curtailed the use of long centre pivots on Vertisols covering 50–100 ha since the high rate of water output by the end gun and low infiltration rate of Vertisols result in ponding and also expose the soil to erosion and compaction.

In addition to the above problems mentioned by Loveday (1984), an ideal irrigation strategy for Zimbabwe Vertisols needs to be worked out and comparisons must be made between frequent light irrigation and infrequent heavy irrigation. Experience on HVE has shown that it is difficult to apply less than 40 mm on basalt using furrow irrigation. Although the normal practice in the industry is to deplete to 50% of available water capacity (AWC) for all soils, Vertisols might benefit from a 75% depletion of AWC and a 25%

replenishment. The practicalities of such a strategy need investigation. The relationship between high rainfall and yields also needs investigation. It has been noted on HVE that a relatively dry summer with adequate irrigation water results in higher cane yields on Vertisols than a relatively wet summer (especially from January to March). The low yields could be due to waterlogging during the very wet summers.

New approaches to soil structure management on Vertisols are required. Hillel (1980) conceives a field planted to row crops as consisting of two distinctly different zones:

1. A planting zone, where conditions are to be optimal for sowing, germination and seedling establishment.
2. A management zone in the inter-row, where soil structure is to be coarse and open allowing maximal intake of water and air, and minimal erosion and weed infestation.

These two zones differ in function as well as in mode of preparation and management. A third zone or function must be considered for the passage over the field by tractors and implements. Random traffic causes compaction all over the field, an effect that can be minimized if travel is confined to special tracks and travel over seedling and water-management zones is avoided. The three zones can be established for a perennial sugarcane cropping programme. Once established, these zones can be maintained consistently over a period of years.

Consequently, it is important to manage both the surface and soil water. There is a need for thorough knowledge of the particular Vertisol and working 'with' rather than 'against' the soil. In Vertisols with poor (coarse) surface structure, tilth and aeration can be improved by maximizing the area of surface wetness by direct flood irrigation and increasing the area wetness by capillarity. Under sugarcane, the natural self-churning property of Vertisols that is essential for the regeneration of soil structure has been impossible because the soil profile is always wet except for a short duration at dry-off. Correct irrigation scheduling is vital and total available moistures should be determined in the field, as discussed previously. Use of instruments such as the neutron probe and tensiometers allows accurate monitoring of soil-water status and improves overall water management (Lanser, 1998). These instruments are being adopted increasingly by both large estates and private commercial farmers in Zimbabwe but sadly are too costly for smallholder farmers.

Technologies such as laser levelling (Besa, 1998) allow precision gradients to be prepared on lands, thus reducing the risks of uneven water applications. HVE is seriously considering purchasing laser levelling equipment in the near future. This again improves the overall water management and lessens the risks of poor crop yields in either under- or over-irrigated sections of lands as well as reducing tail-end water losses. Maintenance of surface drains is also vital to ensure that excess water is removed from land and salinity is minimized.

Irrigation management can be improved with new technologies such as Laeflat irrigation, which uses portable plastic tubing to convey water to the field edge (Siwanja, 1998). This technology reduces excessive water losses that can occur when unlined canals/furrows are used to move water around the fields and thus reduces the likelihood of rising water-tables. This system should also improve the traffickability around the fields, which is often a problem on wet Vertisols.

Research Needs and Opportunities

A large proportion of the Vertisol area in Zimbabwe is being used, but the productivity under rainfed conditions is low. Yields per unit area and total production can be increased by irrigation and opening more land for irrigation. At present there is potential to expand Chisumbanje Estate with a further 40,000 ha. However, irrigation development is very expensive, hence a large proportion of the Vertisol area will continue to be rainfed. There is a need to come up with sustainable technologies to improve productivity under rainfed conditions and the tied furrow technique has proved to be one such technology. The continued non-use of fertilizers by dryland farmers has to be monitored. Under irrigation there might be a tendency to over-exploit the good aspects of Vertisols, hence the present production practices must be monitored and their beneficial effects evaluated to ensure that they are sustainable. There is a need to put more research effort into irrigation management. There has been little attention to improving or using the present natural pastures in a sustainable manner and this should be addressed.

References

Besa, L. (1998) Laser levelling at Nakambala Estate. In: Clowes, M.St.J. (ed.) *Proceedings from the 4th Sugar Seminar*. Zimbabwe Sugar Association Experiment Station, Chiredzi, Zimbabwe, pp. 22–26.

Clowes, M.St.J. and Breakwell, W.L. (1998) *Zimbabwe Sugarcane Production Manual*. Cannon Press, Harare, Zimbabwe.

Hillel, D. (1980) (ed.) *Applications of Soil Physics*. Academic Press, London, pp. 234–241.

Hussein, J. (1981) An investigation into some soil water relationships of some irrigated Vertisols derived from basalt in Zimbabwe. MSc thesis, University of Stellenbosch, South Africa.

Hussein, J., Adey, M.A. and Elwell, H.A. (1992) Irrigation and dryland cultivation effects on the surface properties and erodibility of a Zimbabwe Vertisol. *Soil Use and Management* 8, 97–103.

Jones, E., Nyamudeza, P. and Nyati, C. (1987) Rainfed cropping in Natural Region 5. In: *Cropping in Semi-arid Areas of Zimbabwe.* Department of Agricultural Extension and Technical Services, Harare, Zimbabwe, pp. 531–542.

Kanyanda, C.W. (1989) Properties of some Vertisols of Zimbabwe. In: *Vertisol Management in Africa.* IBSRAM Proceedings No. 9, IBSRAM, Bangkok, Thailand, pp. 99–117.

Lanser, A. (1998) The neutron probe for monitoring soil water and scheduling irrigation – [includes data from the Zimbabwe sugar cane industry]. In: Clowes, M.St.J. (ed.) *Proceedings from the 4th Sugar Seminar.* Zimbabwe Sugar Association Experiment Station, Chiredzi, Zimbabwe, pp. 101–117.

Loveday, J. (1984) Management of Vertisols under irrigated agriculture. In: McGarity, J.W., Hoult, E.H. and So, H.B. (eds) *The Properties and Utilization of Cracking Clay Soils.* Reviews in Rural Science 5. University of New England, Armidale, pp. 269–277.

Low, D., Gibson, W.P., Godwin, R.G.H. and Virgo, K.J. (1984) Variations in characteristics of soils developed on basalt plain in the Zimbabwe Lowveld. *Zimbabwe Agriculture Journal* 81, 121–126.

McCowan, R.L., Keating, B.A., Probert, M.E. and Jones, R.K. (1992) Strategies for sustainable crop production in semi-arid Africa. *Outlook on Agriculture* 21, 21–31.

Mullins, J.A., Webb, A.A. and Coughlan, K.J. (1987) Approaches to erosion control on Vertisols in Queensland, Australia. In: *Management of Vertisols under Semi-arid Conditions.* IBSRAM Proceedings No. 6, IBSRAM, Bangkok, Thailand, pp. 285–297.

Nyamapfene, K.W. (1984) Transmission electron microsopy and electron diffraction studies on the clay fractions of three Zimbabwean Vertisols derived from basalt. *Zimbabwe Journal of Agricultural Research* 22, 111–117.

Nyamapfene, K. (1991) *The Soils of Zimbabwe.* Nehanda Publishers, Harare.

Nyamudeza, P. (1998) Water and fertility management for crop production in semi-arid Zimbabwe. PhD thesis, University of Nottingham, UK.

Nyamudeza, P., Mazhangara, E. and Kamba, E. (1992) Adoption of the tied-furrow technique, and the effects of the technique and previous crop on residual moisture and yields of sorghum and maize on Vertisols. In: Pushparajah, E. and Elliott, C.R. (eds) *Report of the 1992 Annual Review Meeting on AFRICALAND Management of Vertisols in Africa.* Network Document No. 3, IBSRAM, Bangkok, Thailand, pp. 69–81.

Nyati, C.T. (1998) Fertilization. In: Clowes, M.St.J. and Breakwell, W.L. (eds) *Zimbabwe Sugar Cane Production Manual.* Zimbabwe Sugar Association Experiment Station, Chiredzi, Zimbabwe, pp. 109–113.

Probert, M.E., Fergus, I.F., McGarry, D., Thompson, C.H. and Russell, J.S. (1987) *The Properties and Management of Vertisols.* CAB International, Wallingford, UK, 36 pp.

Pushparajah, E. (1991) Success story for Zimbabwe Vertisols. *IBSRAM Newsletter* No. 20, p. 5.

Ristori, G.G., Sparvolie, E., de Nobili, M. and D'Aqui, L.P. (1992) Characterization of organic matter in particle size fractions of Vertisols. *Geoderma* 54, 295–305.

Siwanja, A. (1998) Laeflat irrigation Mkwasine. In: Clowes, M.St.J. (ed.) *Proceedings from the 4th Sugar Seminar.* Zimbabwe Sugar Association Experiment Station, Chiredzi, Zimbabwe, pp. 66–71.

Soil Survey Staff (1975) *Soil Taxonomy.* Agricultural Handbook No. 436. United States Department of Agriculture, Washington, DC.

Thompson, J.G. and Purves, W.D. (1978) *A Guide to the Soils of Rhodesia*. Rhodesia Agriculture Journal, Technical Handbook No. 3, Government Printers, Harare, Zimbabwe, 66 pp.
Vincent, V. and Thomas, R.G. (1960) *An Agricultural Survey of Southern Rhodesia*, Part I. Agro-ecological Survey, Government Printers, Harare, Zimbabwe.

Improving the Productivity of Vertisols for Smallholders on the Accra Plains of Ghana

11

E.K. ASIEDU,[1] Y. AHENKORAH,[2] P. DRECHSEL[3] AND J.W. OTENG[4]

[1]*Soil Research Institute, Academy Post Office, Kwadaso-Kumasi, Ghana;* [2]*Department of Soil Science, University of Ghana, Legon, Ghana;* [3]*IWMI (formerly IBSRAM) Regional Office for Africa, c/o University of Science and Technology, Kumasi, Ghana;* [4]*Agricultural Research Station, University of Ghana, Kpong, Ghana*

Key words: West Africa, Ghana, Vertisols, productivity, farmers, socio-economics, fertilizers

Introduction

Geographical location, soils and vegetation

Vertisols in Ghana are almost entirely confined to the Coastal Savannah (the Accra Plains) and the Interior Savannah zones. They occupy an estimated area of 183,000 ha, of which the bulk of 163,000 ha is located on the Accra Plains (Brammer, 1967). Vertisols of the Accra Plains fall between latitudes 6° and 6°10′ north and longitudes 0°05′ and 0°15′ east (Fig. 11.1). Some of the soils of the area were developed from locally transported weathered gneiss and schist while others were derived from local alluvium. The soils generally occur on very gently undulating land with an elevation range from zero to about 50 m above sea level. According to Adu (1989), the general slope of the land on which they occur is about 1–2%, rarely exceeding 5%, and they usually occupy the whole topography from low-lying summits to valley bottoms. The soils occur under a rainfall of 900–1400 mm, which is spread over two rainy seasons per year. Typical soil characteristics of the Vertisols on the Accra Plains are shown in Tables 11.1 and 11.2.

(eds J.K. Syers, F.W.T. Penning de Vries and P. Nyamudeza)

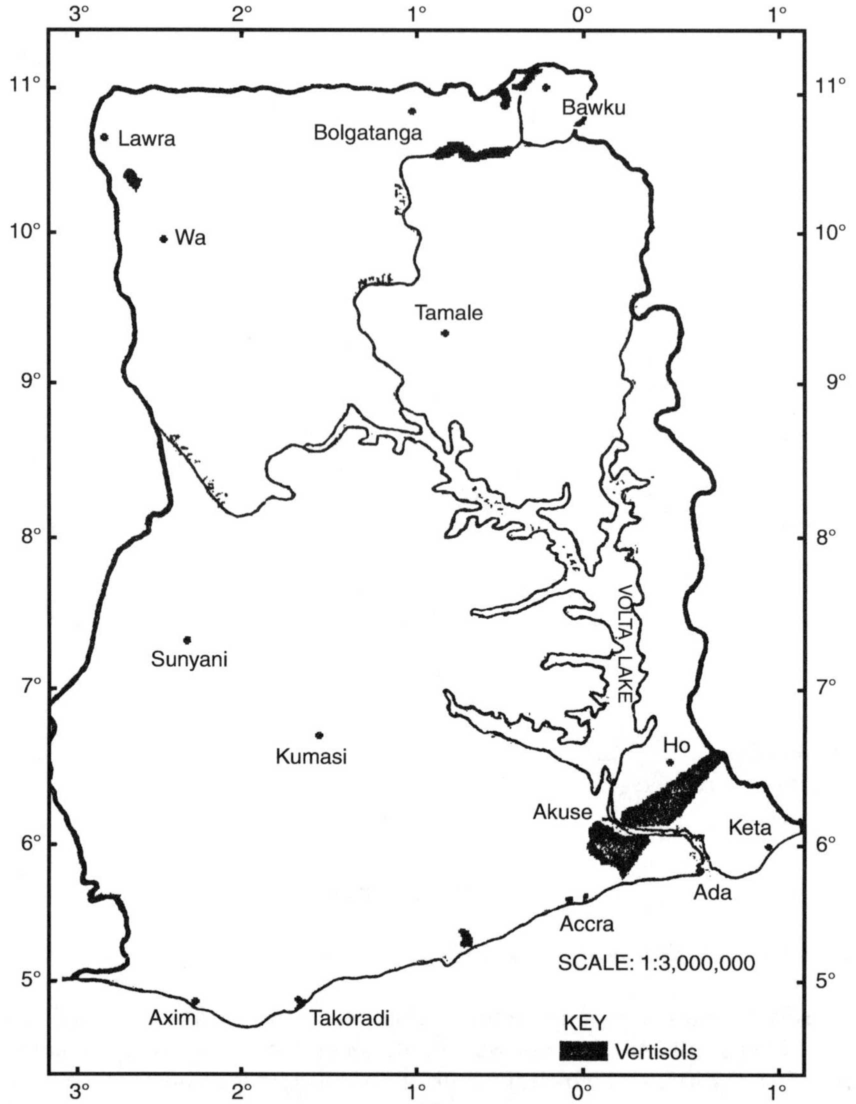

Fig. 11.1. Map of Ghana showing the geographical locations of the Vertisols.

The vegetation of the area is savannah grassland, the greater part of which is covered with medium to tall grasses. The original vegetation as recorded by Brammer (1955) showed *Vetiveria fulvibarbis* as the dominant grass species with very frequent *Brachiaria falcifera*. Other species recorded by Brammer (1955) are *Schzachyrium semiberne*, *Euclasta condylotrica*, *Andropogon canaliculatus*, *Andropogon gayanus*, *Sporobolus pyramidalis*, *Heteropogon contortus*, *Imperata cylindrica*, *Cymbopogon giganteus*, *Cassia mimosoides* and *Bambusa* spp. The most common legumes are *Tephrosia elegans* and *Cassia mimosoides*.

Table 11.1. Analytical data on physical properties of the Akuse series (ARS, Kpong). (Source: Amatekpor and Dowuona, 1995.)

Horizon	Depth (cm)	Bulk density (mg m^{-3})	Moisture content (cm^3 cm^{-3})	Air-filled porosity (%)	Particle size distribution (%)			
					Sand	Silt	Clay	Texture
Aul	0–7	1.4	0.20	27.17	48.1	5.5	46.4	Sandy clay
Au2	7–21	1.6	0.15	24.62	46.0	5.2	48.8	Clay
Bw	21–43	1.6	0.21	18.62	40.7	2.6	56.7	Clay
Bkss	43–66	1.6	0.18	21.62	38.5	2.6	58.9	Clay
BCk	66–90	1.6	0.11	13.53	44.5	2.5	53.0	Clay

The dominant tree species comprise *Combretum ghasalense, Anona senegalensis* and *Ceiba pentandra*. Other frequently occurring species are *Bauhinia thonningii, Pterocarpus erinaceus* and *Bridelia ferruginea.*

Agricultural importance

Smallholder crop production constitutes the major farming activity on the Vertisols of the Accra Plains. The crops are cultivated, usually under rainfed conditions, as mono-crops or in mixed-cropping systems. Crop production brings cash income to 88.6% of the farmers. Most of the farmers are smallholders, whose farm sizes range between 0.2 and 2.0 ha, with the majority of the farmers cultivating between 0.4 and 0.8 ha (Kwadzo, 1995). The major crops cultivated in the area are maize, rice, cassava, cowpea, groundnuts, sugarcane, yam, sweet potato, plantains and such vegetables as okra, pepper, garden eggs (aubergines), cabbage and tomato.

Only a few commercial poultry and livestock farms are found in the area, though most households have a small flock of fowls, sheep and goats. Cattle rearing is increasing, with some farmers keeping a sizeable number of beef cattle to take advantage of the extensive grassland of the area.

Even though Vertisols are potentially productive, the full potential of the soil is not attained. For example, average grain yields of 0.660 t ha^{-1} maize are obtained by farmers in the area as against potential values of 4.20 t ha^{-1} (Ahenkorah *et al.*, 1995).

Despite the fact that the area occupied by the Vertisols of the Accra Plains is small (only 183,000 ha) compared with the total area of Ghana (23,946,000 ha), these soils play a very important role in the socio-economic activities of the country because of their strategic location. The Accra Plains Vertisols are quite close to the major cities of Accra and Tema and therefore in the peri-urban zone of these cities, producing food for an estimated population of over 2 million inhabitants of Accra, Tema and their environs. In addition, the soils provide a livelihood for over 50% of about 700,000 people who live on them (Kwadzo, 1995).

Table 11.2. Analytical data on chemical properties of the Akuse series (ARS, Kpong). (Source: Amatekpor and Dowuona, 1995.)

Horizon	Depth (cm)	pH 1.1 (H_2O)	pH 1.2 ($CaCl_2$)	Electrical conductivity (dS m^{-1})	Organic carbon (%)	Extractable bases (cmol(+) kg^{-1}) Ca	Mg	Na	K	Bases	Cation exchange capacity (cmol(+) kg^{-1})	Base saturation (%)	Total phosphorus (mg kg^{-1})	Available phosphorus (mg kg^{-1})	Nitrogen (%)
Aul	0–7	6.5	5.9	0.01	1.03	22.2	8.2	0.70	0.23	31.33	34.32	91.3	2.9.0	1.75	0.08
Au2	7–21	6.9	6.6	0.01	1.03	23.0	8.2	0.83	0.09	32.12	37.24	87.3	117.6	1.58	0.04
Bw	21–43	7.3	6.2	0.01	1.02	25.8	9.8	1.83	0.08	37.51	45.90	81.7	116.4	0.87	0.03
Bkss	43–66	7.7	7.5	0.02	0.65	27.2	11.0	2.5	0.08	40.83	49.80	82.0	118.8	0.95	0.03
BCk	66–90	8.2	7.6	0.11	0.46	36.6	10.8	3.0	0.08	50.58	58.48	96.2	178.0	1.49	0.02

Socio-economic characteristics

The household size of the farmers is reported to range between one and 19 persons with an average size of seven. The population is made up of 46.7% children aged between 0 and 18 years and 53.3% adults aged over 18 years. Of the adult population, at least 76.1% undertake some form of farming activity, on either a full-time or a part-time basis. Males constitute a slightly larger group and form 51.3% while females constitute 48.7% of the households of the people who work on the Accra Plains Vertisols. This might be explained by the fact that many of the females aged between 15 and 35 years, who form the bulk of the active population (1984 census), have migrated to the nearby Accra/Tema metropolis.

The next most important occupation in the area, after crop farming, is trading in farm produce, which is predominantly carried out by the women. The major types of labour used in agricultural production are family labour (males, females, children and dependants) and hired labour. The men perform the more tedious work of land clearing, seedbed preparation and weeding, while female labour and, to a lesser extent, children carry out the jobs of planting, harvesting and carting.

Other sources of income of the farmers in the area are salaried work and such activities as charcoal burning, carpentry, shoe mending, fishing, gin brewing, fruit collecting (particularly mangoes), hunting, vehicle repairs, masonry, oyster shell collecting, tailoring, wine tapping (the tapping of palmwine, a local sweet alcoholic beverage from palm trees), etc. It is estimated that the average annual income of the farmers is ¢1,969,266 or $1660.43 (Kwadzo, 1995).

The high productive potential of these soils and their ability to support many different crop and livestock populations, if properly harnessed, has the ability to counteract the migration of the youth to the urban areas of Accra and Tema in search of non-existent white-collar jobs.

Recent Interventions for Improving the Productivity of Vertisols of the Accra Plains

Landform technologies

One major limitation to the use of Vertisols of the Accra Plains is ponding or surface waterlogging during the rainy seasons. In an effort to find a solution to this problem, the Ghana/IBSRAM Vertisols Project, after a series of field experimentations, has developed landform technologies that can offset this problem. According to Duah-Yentumi *et al.* (1992) and Ahenkorah *et al.* (1995), the cambered bed offers a ready solution to this problem of poor surface drainage and gives the highest grain yield for maize and cowpea, followed by the Ethiopian bed, the ridge and the flat bed in that order (Table 11.3).

Asiedu *et al.* (1997a), evaluating these landforms on the basis of soil physical parameters, also confirmed the superiority of the cambered bed to the other landform technologies. According to their work, total porosity and air-filled porosity were highest and therefore most desirable in the cambered bed. Bulk density was lowest in the cambered bed and highest in the flat bed (Table 11.4).

Results from a parallel work by Asiedu *et al.* (1997b) also indicated that infiltration rate and sorptivity were highest in the cambered bed and lowest in the flat bed. Work done by Asiedu and Bonsu (1997) on analysis of the drying efficacy of the landform technologies shows that the time to incipient ponding of the cambered bed is longer than that of the other landforms. This suggests that the problem of ponding is minimal with the cambered bed when compared with the other landforms. These landform technologies have been introduced to farmers through on-farm trials (Table 11.5) and an open/field day exposition. The initial awareness of the technology was created among 85 farmers and eight Ministry of Food and Agriculture (MOFA) extension staff in the area. The technology is yet to be adopted extensively. According to a survey conducted by Kwadzo (1995), only 3% of the farmers of the area use the cambered bed while as many as 63% of the farmers continue to plant on the

Table 11.3. Mean grain yields (t ha^{-1}) of maize and cowpea over a 3-year period (1992–1994) for four landform technologies. (Source: Ahenkorah *et al.*, 1995.)

	Mean grain yield (t ha^{-1})	
Landform	Maize	Cowpea
Cambered bed	1.21^{a}	0.51^{a}
Ethiopian bed	1.15^{a}	0.49^{ab}
Ridge	0.84^{ab}	0.36^{bc}
Flat bed	0.70^{b}	0.30^{c}
LSD	0.3076	0.1332

Means followed by same letters in a column are not significantly different at 5%.

Table 11.4. Some soil physical parameters for four different landform technologies on the Vertisols at the Agricultural Research Station (ARS), Kpong. (Source: Asiedu *et al.*, 1997a.)

Landform	Bulk density (g cm^{-3})	Total porosity (%)	Air-filled porosity (%)
Flat bed	1.466^{a}	41.0^{b}	0.17^{c}
Ethiopian bed	1.412^{ab}	42.4^{b}	7.10^{bc}
Ridge	1.419^{ab}	42.2^{b}	11.40^{ab}
Cambered bed	1.354^{b}	45.3^{a}	17.10^{a}

Means followed by same letters in a column are not significantly different at 5%.

Table 11.5. Effect of landform on mean grain yield of maize from on-station and on-farm trials (1993–1995). (Source: Ahenkorah *et al.*, 1995.)

	Grain yield of maize (t ha^{-1})						
		On farm					
Landform	On station	A	B	C	D	E	Mean
Cambered bed	2.49	3.80	1.80	2.14	2.97	1.84	2.51
Flat bed	1.33	1.50	1.06	1.05	0.77	1.84	1.04

flat and 15% on ridges. The reasons are discussed in subsequent pages. Asiedu (1998), however, observed that the flat bed is superior to the other landform technologies when water is limiting, such as in the dry season.

Tractor/power tiller services and tillage timing

Generally, Vertisol farmers use hoes, cutlasses, axes, mattocks, rakes, forks, power tillers and tractors. Cutlasses and hoes are basic tools used by almost all the farmers. About 80% of farmers on the Vertisols use tractor services basically for land preparation while another 25% use power tillers, but these are not owned by the farmers themselves. They have to pay for such services.

The farmers believe that tractor/power tiller services ease working the Vertisols, open up the soil, enable faster and better land preparation and contribute to higher yields. However, the reasons given by the farmers (21%) for not using tractor services are the high cost of such services and the fact that their land holdings or farm sizes are too small for tractor use.

Work done by Antwi and Asiamah (1996) shows that soil moisture content of 24–33% is the range most appropriate for the use of implements and machinery on the Vertisols of the Accra Plains. Even though farmers do not take measurements of the moisture status of the soils before using implements and machinery on them, farmers are able, through experience, to determine at what moisture content the Vertisols are most suitable for tillage. Consequently, in the Accra Plains, most farmers consider the end of the minor rainy season as the most suitable time for tilling the land for the ensuing cropping season, while others till the land at the onset of the major season rains when the soil is neither too wet nor too dry.

Use of improved seeds

Improved seeds include clean and viable seeds of superior varieties of crops that give higher yields and show resistance to drought, waterlogging, pests and diseases. In Ghana, such seeds are sold to farmers through either the Ministry of Food and Agriculture (MOFA) or private merchants. Improved seeds are often available and Vertisol farmers may not have to travel long distances (usually not more than 5 km) to obtain them.

Available statistics show that about 60% of the Vertisol farmers use improved seeds. About 40% of the farmers do not use improved seeds even though they are aware of the advantages of the use of such seeds. Their reasons for not using improved seeds include the high cost of such seeds, lack of financial resources to purchase them and the general unavailability of such seeds. While some of the farmers are satisfied with their own seeds, others believe that the produce from improved seeds does not taste normal.

Kwadzo (1995) reports that the most important factor that determines the use of improved seeds by the farmers is the availability of tractor and power tiller services. This is followed by the price : cost ratio and financial resources. Where tractor services are available for land preparation, work is timely and the soil is well turned. The farmers are convinced that this will bring out the full potential of improved seeds and, therefore, under such conditions, will use them.

Chemical weed control

One of the major problems facing farmers on the Vertisols in Ghana is weed control. Due to the high moisture retention of the soils, weeds establish and grow very fast. Weeding with the traditional hoe and cutlass becomes difficult in the cropping season since the soil sticks to the implements tenaciously. The use of herbicides in the control of weeds on these soils has proved very promising. Preliminary trials under the Ghana/IBSRAM Vertisol project by Ahenkorah *et al.* (1995) were aimed at evaluating and selecting the effective rate of application of Roundup and Touchdown (based on the active ingredient glyphosate) for the control of the most common perennial weeds, *Imperata, Andropogon* and *Cyperus* spp. on the Vertisol at Kpong. The results indicated that a dose of 3–4 l ha^{-1} (1.4–2.0 ai ha^{-1}) appeared more effective than disc ploughing and harrowing (conventional tillage) in controlling the weeds (Table 11.6).

Table 11.6. Effect of weed control treatments on weed densities. (Source: Ahenkorah *et al.*, 1995.)

	Weed count (no. m^{-2})			
Treatment	Grasses	Broad leaves	Sedges	Mean
Herbicide	10.5	10.9	9.5	10.3
Tillage	20.1	19.9	11.7	17.2
Slash-and-burn	14.7	16.6	7.7	13.0

A greater number of Vertisol farmers in Ghana do not use herbicides for weed control even though most of the farmers are aware that the use of herbicides speeds up weeding operations on their farm. It is reported that only 30.5% of the farmers use herbicides. The reasons given by the farmers for not using herbicides are the high cost of the herbicide and the lack of financial resources. Other reasons are that herbicides are not available all the time and the fear that misapplication of the chemical may negatively affect their food crops. Interestingly, most of the farmers who use herbicides for weed control do not use the recommended rates but use lower rates because of the high cost. A more detailed study involving cost–benefit analysis is very necessary for future work.

Use of pesticides (insecticides, fungicides, nematicides and rodenticides)

The control of insects and other pests is very important for productive crop husbandry on the Vertisols of the Accra Plains and farmers are aware of this. It has been observed that, under waterlogged conditions on Vertisols, fungal diseases are prevalent. Similarly insect damage to crops, especially the vegetables cultivated on the fertilized Vertisols, is common. Rodent attacks on crops, particularly grains, are also very common. These pests seriously affect crop production and reduce farmers' income. Available statistics indicate that about 39% of the farmers (on the Vertisols) are reported to use pesticides (fungicides, rodenticides, insecticides and nematicides) while over 60% of the farmers do not use any pesticides at all (Kwadzo, 1995). Their reasons for not using pesticides are similar to those for not using herbicides.

No work has been done in this area on the Vertisols of the Accra Plains. The effective control of these pests by the use of pesticides and other methods needs to be investigated thoroughly.

Improved cropping systems

Farmers on the Vertisols of the Accra Plains practise different forms of cropping systems. Mono-cropping, for most of the crops grown in the area, is the commonest cropping system, with about 39% of the farmers practising it. Other farming systems practised by the farmers include mixed cropping, which is undertaken by 13.6% of the farmers; multiple cropping, which is undertaken by 19.7%; and mixed farming (crops and livestock). The farmers often practise more than one farming system.

However, because of differences in rooting and feeding behaviours and the ability of legumes to fix nitrogen biologically to the soil, legume–cereal in rotation or in mixed-cropping has been recommended for the cropping systems on Vertisols. This practice has caught on well with the farmers and available statistics indicate that as many as 80% of the farmers in the area practise legume–cereal in rotation or in mixed cropping. Farmers say that this practice increases yield, fixes nitrogen and prevents pest infestation. A few farmers in the area do not practise this recommendation because some of them are not used to the system or have no knowledge of its benefits or lack the skills for its successful practice. Others do not practise it because they believe it brings less income and, for a few others, land is sufficiently available for mono-cropping. Farmers who produce only vegetables or irrigated rice are less interested in this system.

Soil fertility management

Generally, farmers in the area are aware of the need for proper management of soil fertility for increased crop yield. Land fallowing has been the most common method of improving productivity of the Accra Plains Vertisols. However, with increasing pressure on land and consequent reduction of the fallow period, this practice does not hold any promise of sustainability as the typical fallow vegetation consists of grasses.

Preliminary results of ongoing work by Asiedu *et al.* indicate that the use of such leguminous cover crops as *Mucuna* and *Stylosanthes* holds great promise in reducing the fallow period and at the same time improving the productivity of these soils. This technology has not been introduced to farmers in the area for adoption yet, as the work is ongoing.

Other methods that several farmers have adopted include manuring, mulching, bunding or a combination of these methods.

The use of manure is not very popular among the farmers. On the whole only 18.2% of the farmers use manure on their farms. The types of manure used are cow dung, which is used by 7.6% of the farmers, poultry manure, used by 6.9% of the farmers, and bat guano. The farmers who took part in the survey were not able to estimate the quantities of manure they use. Most farmers in the area (over 80%) do not use manure on their farms because it is unavailable as livestock keeping does not form part of the culture of the local people. High transport cost, lack of knowledge about its use and the belief that its spreading is time-consuming and tedious are other reasons. Most of the farmers also do not use manure for cultural/traditional reasons and because they are satisfied with the fertility status of their land. They also believe that the use of manure promotes weed growth. Little work on the use of manure has been done but there is a need for such studies.

Fertilizer application

The beneficial effects of chemical fertilizers in improving soil productivity are known to farmers. Even though the Vertisols are potentially fertile and productive, the soils have been cropped by the local farmers for centuries with very little input. Amendment of such soils with mineral fertilizers should therefore be beneficial for increased crop production.

Results of work done by Ahenkorah *et al.* (1995) indicated that the application of chemical fertilizers resulted in significant increases in grain yields on the Vertisols of the Accra Plains of Ghana (Table 11.7). This was more pronounced on the cambered bed. According to the results of a socio-economic survey, about 60% of farmers on the Vertisols use fertilizers. About 40% of farmers do not use fertilizers for the following reasons: fertilizers are costly, farmers do not have the financial resources to purchase them, irregular availability of fertilizers, the land is already fertile, bush fallowing is sufficient for fertility regeneration, and the farmers' belief that fertilizers change the taste and quality of their produce (Kwadzo, 1995).

Social and Economic Factors affecting Vertisol Productivity in the Accra Plains

It can be observed from the presentation made so far that there are many improved technologies, which when adopted by smallholders can improve the productivity of these soils. There are equally many social and economic factors that militate against the adoption of these improved technologies and for which solutions have to be sought. Some of these factors can be summarized as follows.

Lack of finance/credit

Information gathered from a survey of the area indicates that the major problem of the farmers is lack of credit (more than 50% of the farmers do not have access to credit). The traditional sources are the market women, professional

Table 11.7. Summary of maize grain yield for three levels of fertilizer on four landforms. (Source: Ahenkorah *et al.*, 1995.)

	% Recommended fertilizer rates			
Landforms	0	50	100	Means
Flat	0.700^{b}	1.387^{c}	2.422^{b}	1.503^{c}
Ridges	0.842^{ab}	1.867^{bc}	2.960^{ab}	1.890^{b}
Ethiopian	1.149^{a}	2.057^{b}	3.114^{a}	2.107^{b}
Cambered	1.213^{a}	2.767^{a}	3.458^{a}	2.479^{a}
Means	0.976^{c}	2.019^{b}	2.988	

Means followed by the same letters in a column are not significantly different at 1%.

money-lenders, the commercial banks, family members and, in a few instances, the local churches. Farmers who succeed in getting credit from any of these sources often make payment either with farm produce at farmgate price or pay cash at high interest rates.

A planned credit system, when introduced in the area, can provide much needed finance for farmers to purchase inputs. As suggested by Kwadzo (1995), this should be started with a selected number of farmers and crops (e.g. pepper, rice, maize) to assess their impact and performance in terms of repayment before it is applied to the entire area for all interested farmers.

Lack of tractor services and other inputs

Farmers in the area recognize the difficult nature of the Vertisols and consider the tractor as the best vehicle for effective land preparation. In fact, the adoption of most of the improved tillage systems by the farmers depends on whether they have access to tractor services or not. Unfortunately, this service is not always available and where such service is available many of the farmers cannot afford the high cost involved. This problem could be solved if a centrally placed, or ideally several, tractor-service centre(s) are established with adequate maintenance and training support units. Such service could be rendered to the farmers on credit and payment demanded after harvest. The Agricultural Engineering Services Department of MOFA could render such a service.

The problem of year-round availability of other inputs, such as fertilizers, improved seeds, agrochemicals, etc., could be solved if sale centres that do not run out of stock are established in all the towns of the Vertisols area. The inputs could be sold to farmers on credit and the money recovered after harvest. A commercial unit of MOFA could be created to handle this effectively.

Labour constraints

Labour supply in the area is seasonal and becomes scarce during the peak farming activities when it is needed most. During school holidays, students offer their labour for income to support their parents. Labour therefore becomes readily available during school breaks. The nature of the work to be done also determines labour availability. For example, labour is more readily available for harvesting than for planting since some food would be obtained in the former activity for immediate consumption. It is also more difficult to attract labour to undertake tedious activities such as tillage and weeding.

Considering the importance of the timeliness of farming operations on Vertisols, the lack of labour at critical periods can cause a major loss in productivity. To overcome this problem, farmers must be encouraged to use labour-saving devices such as tractors, herbicides, etc. This can be achieved through extensive education during the farmer-training programmes suggested below and through services provided by the tractor service centre(s) suggested above.

Agricultural extension system

The major source of technical advice for Vertisol farmers is the extension services provided by the MOFA. They reach over 70% of the farmers in the area. Other sources of extension education include the Agricultural Research Station (ARS), Kpong (6%), church groups (1%), non-governmental organizations (NGOs) (2%) and others (15%). This underscores the need to involve the front-line extension staff of MOFA in any 'technology transfer' programme for farmers. The efficiency of the extension system in diffusing innovations on the efficient management of Vertisols will therefore influence the adoption of improved practices on these soils. The following observations can be made about the agricultural extension system in the area.

Lack of efficient research–extension linkages

In a few situations, attempts are made by research scientists in the area to transfer technologies and improved practices to the farmers directly. Such an effort is not likely to achieve the desired impact. This certainly accounts for the low adoption of improved practices, such as the cambered bed technology.

High farmer : extension agent ratio

As in other parts of Ghana, there is a high farmer : extension agent ratio in the area such that one extension agent is responsible for 1500 farmers. This makes it difficult for effective extension work and limits the diffusion and adoption of technologies.

Extension methods

Most of the extension agents use individual, farmer-to-farmer teaching methods instead of using the group approach in the diffusion of improved technologies. This is a serious limitation that has to be addressed.

High cost of inputs/services and lack of subsidies

The ability to pay for services and other inputs by the farmers is a major determinant of the adoption of improved technologies and practices. Most of the Vertisol farmers, who are smallholders and resource-poor, cannot afford the high cost of tractor services, improved seeds, fertilizers, agrochemicals or even labour. This problem is aggravated by the lack of subsidies for these inputs. The establishment of a credit scheme proposed above will go a long way to give the farmers the needed resources to purchase these inputs.

Lack of knowledge/skills

Some of the farmers do not adopt some of the improved practices and technologies because they have no knowledge or lack the requisite skill to practise them efficiently. For example, the cambered bed technology has not been adopted by many of the farmers simply because the local tractor operators are not familiar with the land-shaping technique for the preparation of the cambered bed. Some farmers do not practise legume–cereal rotation because they have no knowledge of the benefit of the system. Other farmers do not use fertilizers and agrochemicals due to lack of knowledge and skill in their use.

Farmer-training programmes and more participatory/on-farm trials to expose farmers to improved practices for the management of Vertisols could provide a solution to this problem. This training could be conducted at the ARS, Kpong, where facilities already exist.

Land tenure

Land in the area may be acquired through freehold tenure, cash renting, inheritance, gift or share cropping. Customarily, all land belongs to the 'stool', which is the traditional authority. The land is sub-owned by families and its use is controlled by the family head, though members of a family have right of use of the family land. It is significant that about 33% of the farmers acquired their lands either through renting or share cropping. This is a considerable disincentive to farmers, who do not have the security to develop such lands for long-term use.

Government policy intervention has been seen as the only possible solution to land tenure problems in the country. However, because of the traditional links between land and families, attempts in the past by governments to intervene in land issues have not been successful. While the 'stool' stands for traditional authority in the southern part of Ghana where the Accra Plains are located, the symbol of traditional authority in the northern sector of Ghana is the 'skin'.

Cultural or traditional beliefs/taboos

There are many taboos in the area. The major taboo affecting agricultural activity is that which prohibits the inhabitants from going to farm and/or to work with hoes on designated days. These 'rest days' are reserved to honour the gods of the land and war victims. It is believed that, when this is disobeyed, the rains will fail. Some of the communities have Mondays while others have Mondays and Thursdays as taboo days. Given that Sundays are already non-working days, the implication is that some of the communities have only four working days in a week.

If the many churches in the area have not succeeded in changing the situation, the probability that any scientific intervention can bring any change to the situation is very remote. The best that can be done will be to develop a workplan such that off-farm activities, such as buying inputs, preparing seeds and planting materials, etc., are done on the taboo days.

Farmers' perceptions

Farmers, like researchers, have certain beliefs and perceptions that prevent them from adopting certain improved practices. Some farmers, for instance, will not use improved seeds for planting because they believe or have noticed that produce from such seeds does not taste normal. Another group of farmers in the area does not use fertilizers because they believe fertilizers change the taste and quality of the farm produce or favour weed growth. Farmers who do not use manure see it as comparable to human excreta, which they consider abhorrent to touch by hand. This situation could be changed with more participatory and/or demonstration, on-farm trials or through education during the farmer-training programmes suggested above.

Conclusion and Recommendations

There are some technologies and practices that after improvement and adoption can enhance the productivity of the Vertisols of the Accra Plains of Ghana. These include the use of the appropriate landform technologies and fertilizers. Many of the farmers are aware of some of these innovations, yet their adoption is being hindered by a number of factors, which include the lack of tractor services and agrochemicals, the high cost of inputs and farmers' scanty financial resources. Other factors such as land tenure, traditional beliefs of the people, ineffective research–extension linkage and the farmers' own perceptions about certain improved practices also affect the rate of adoption of technologies and improved practices and consequently the productivity of these soils. There are also many information gaps, especially with respect to economic analyses of interventions, which need to be filled through appropriate research programmes.

The following recommendations are made to find solutions to the problems that hinder the successful adoption of improved practices by farmers and to indicate the relevant research needed to fill the information gap for the realization of the full potential of these soils:

1. Since a major problem of the farmers of the area is lack of credit facilities, the introduction of a planned credit system will provide much-needed finance. Such a service can be provided by financial institutions, NGOs and/or the local churches. It should be started with a selected number of farmers to assess impact and performance in terms of repayment, before it is made available generally.

2. Training of extension staff on Vertisol technologies is essential. Since the front-line staff of MOFA are involved directly in the diffusion of agricultural innovations to the farmers, they should be exposed to innovations and actively involved in the transfer of any technology to the farmers. To make this work effectively, there is a need to strengthen the research–extension–farmer linkage in the area.

3. A short group training programme and participatory on-farm research with the farmers should be instituted in the area to expose the farmers to improved practices, new technologies, etc. and to collaborate with them to improve their performance. Funds for such a programme could be solicited from the District Assemblies, NGOs and donors. The training can be conducted at the ARS, Kpong, where facilities are already available.

4. Sales centres should be established in all the communities that work on Vertisols. Such centres will stock and sell such farm inputs as fertilizers, herbicides, pesticides, improved seeds, etc. to the farmers at all times on their doorsteps. This should be subsidized. A mechanism could be put in place to sell these inputs to needy farmers on credit and the money recovered after the crop harvest.

5. Many of the farmers who work on Vertisols use tractor services, which are very costly and also not always available. The establishment of one or several tractor service centres with adequate maintenance and training support units in the area will go a long way to providing tractor services to the farmers to enhance their farming activity. Such a service could again be rendered to needy farmers on credit and the cost recovered after the crop harvest.

6. Labour availability has been identified as a major constraint in the productive use of the Vertisols of the Accra Plains for farming. Farmers in the area should be encouraged to employ labour-saving methods such as the use of the tractor for land preparation and herbicides for weed control.

7. Solutions to most of the problems identified would be more effective if the farmers are organized into cooperatives or crop associations. However, given the history of cooperative associations in Ghana, sufficient care must be taken and the necessary precautions taken to ensure success.

8. The following areas of research are necessary to improve the productivity of the Vertisols in Ghana and to prevent people from avoiding their cultivation:

- Most of the technologies and improved practices that have been recommended to the farmers do not carry information on the cost-effectiveness of such innovations. Farmers will be more willing to adopt any intervention when they are convinced of its profitability. It is therefore necessary to consider the economic analyses of such practices as fertilizer application, cambered bed technology, use of manure, use of herbicides in weed control, use of tractor services, pesticide use, etc. Information on the economic analyses of such innovations, when made available to the farmers, will assist them in making decisions on technology adoption.

- Though the cambered bed or raised bed technology increases productivity, there is a need to generate information through on-farm trials to determine its longevity/stability and long-term profitability.
- In recognition of the seriousness of pest problems on the Vertisols of the Accra Plains, there is a need for research on effective weed and pest management, in conjunction with the extension agents and farmers.
- Labour availability and the high cost of tractor services constitute a major problem for farmers. There is a need to establish on-farm trials with farmers in collaboration with the extension agents to evaluate the economic potential of various sources of labour including hand, draught animal and tractor power.

It is hoped that recommendations from such studies will fill the existing information gap and remove many of the constraints that confront smallholder farmers who cultivate the Vertisols of the Accra Plains of Ghana.

Acknowledgements

The authors wish to thank Mr E.O. Ampontuah and Mrs Rosalind Ceasar, both of the Soil Research Institute, Ghana, and Mr S. Asante-Mensah, Principal of Bunso Cocoa College and an extension specialist, for their various contributions to the preparation of this chapter.

References

Adu, S.V. (1985) Vertisols and Acrisols in Ghana: their physical and chemical properties and management in Africa. Cited by Acquaye, D.K. and Owusu Bennoah, E. (1989) Chemical characteristics of some Vertisols of Ghana. In: *Vertisol Management in Africa.* IBSRAM Proceedings No. 9, IBSRAM, Bangkok, Thailand, pp. 327–346.

Ahenkorah, Y., Owusu-Bennoah, E., Oteng, J.W., Darkwah, E.O., Johnson, B.K., Dowuona, G.N.N. and Amatepkor, J.K. (1995) *Management of Vertisols for Sustainable Rainfed Smallholder Production in Ghana.* Ghana/IBSRAM Vertisol project phase 1 Final Technical Report. Department of Soil Science, University of Ghana, Legon.

Amatekpor, J.K. and Dowuona, G.N.N. (1995) *Site Characterisation of IBSRAM Vertisol Project.* Department of Soil Science Publication, Technical Report No. 95/001, University of Ghana, Legon.

Antwi, B.O. and Asiamah, R.D. (1996) *Morphological and Hydraulic Properties of the Soils of Kpong Irrigation Project.* Soil Research Institute Technical Report No. 186. SRI, Ghana.

Asiedu, E.K. (1998) Soil and water management on Vertisols of the Accra Plains of Ghana. Paper presented at the Research–Extension–Farmer Linkage Workshop on Land and Water Management and Conservation Practices in Ghana (2–5 February 1998). Soil Research Institute, Kwadaso.

Asiedu, E.K. and Bonsu, M. (1997) Analysis of the drying efficiency of some Vertisol management technologies using a simple model. Paper presented at the 15th AGM of the Soil Science Society of Ghana (24–27 June 1997). KNUST, Kumasi.

Asiedu, E.K., Ahenkorah, Y., Bonsu, M. and Oteng, J.W. (1997a) Assessment of the performance of some Vertisol management technologies using soil physical parameters. Paper presented at the 15th Annual General Meeting of the Soil Science Society of Ghana (24–27 June 1997). KNUST, Kumasi.

Asiedu, E.K., Ahenkorah, Y., Bonsu, M. and Oteng, J.W. (1997b) Infiltration and sorptivity studies on some landform technologies for managing Vertisols. Paper presented at the 15th Annual General Meeting of the Soil Science Society of Ghana (24–27 June 1997). KNUST, Kumasi.

Brammer, H. (1955) *Detailed Survey of the Kpong Pilot Irrigation Area.* Gold Coast Department of Soil and Land Survey, Memoir No. 1, Kumasi, Ghana.

Brammer, H. (1967) *Soils of the Accra Plains.* Soil Research Institute Memoir No. 3. SRI, Ghana.

Duah-Yentumi, S.E., Owusu-Bennoah, E., Oteng, J.W., Acquaye, D.K. and Ahenkorah, Y. (1992) Improving the management of Vertisols for sustainable cropping. Report on the major season trial with maize on Vertisol of the Accra Plains. In: Pushparajah, E. and Elliot, C.R. (eds) *AFRICALAND Management of Vertisols in Africa.* Network Document No. 3, IBSRAM, Bangkok, Thailand, pp. 19–23.

Kwadzo, G.T.M. (1995) *Socio-economic Study of Small-scale Farm Households on the Vertisols of the Accra Plains.* Department of Agricultural Economics and Farm Management, University of Ghana, Legon.

From Plot to Watershed Management: Experience in Farmer Participatory Vertisol Technology Generation and Adoption in Highland Ethiopia

12

M.A. JABBAR,[1] TEKALIGN MAMO[2] AND M.A. MOHAMED SALEEM[1]

[1]*International Livestock Research Institute, Addis Ababa, Ethiopia;*
[2]*Ethiopian Agricultural Research Organization, Addis Ababa, Ethiopia*

Key words: East Africa, Ethiopia, Vertisols, watershed management, farmers, implements, land use

Background

Highlands (1500 m above sea level) cover about 10% of tropical Africa, mostly in East Africa, but they support close to half the human and about 30% of the total livestock populations. Major problems encountered in the highlands are poverty and malnutrition, low crop and livestock productivity, widespread land degradation, and underuse of some resources due to specific constraints.

The highlands of Ethiopia constitute over 60% of the East African highlands and account for more than 80% of the agricultural lands. In addition, they serve as the major catchment for many tributaries that feed the River Nile. The majority of agricultural production is in the hands of the peasant sector, and farm technology has undergone little change for hundreds of years. As in other countries in the region, the smallholder farmers in the Ethiopian Highlands are poor, individual landholdings are no more than 0.5–2.5 ha, family sizes are large and land productivity is low; thus food requirements are not met. Deforestation is common, steep-slope cultivation and overgrazing have become rampant in response to the increasing pressure on land, and fallowing as a means of fertility replenishment has almost disappeared or is ineffective because of the use of dung and crop residues for fuel (which represents an annual loss in crop production of 700,000 t of grain). A major problem for agricultural growth in the country is land degradation, including soil erosion and loss of soil fertility, water resource degradation and loss of biodiversity. At present rates, it is estimated that land degradation could destroy the farmlands of some 10 million highland farmers by 2010 (Sutcliffe, 1993). Land degradation also contributes to the deterioration of grazing land. This in turn has contributed to feed shortage due to the breakdown of traditional systems of livestock grazing and to the low performance of livestock.

While population pressure has induced cultivation and livestock grazing on steep slopes and fragile lands, causing serious vegetation loss and soil erosion, a large area of Vertisols (heavy black clay soils) remain underused. Vertisols cover some 43 million ha comprising 19% of the total land area in sub-Saharan Africa. Nearly 30% of the Vertisols area is located in Ethiopia alone, particularly in the highland region (Mohamed Saleem, 1995). Vertisols are productive soils but difficult to manage due to their poor internal drainage and resultant flooding and waterlogging during the rainy season. Consequently, Vertisols in Ethiopia are largely used for dry season grazing and under 30% of the area is cultivated. The cultivated Vertisols are exposed to soil erosion and give low yields because the fields are ploughed before the main rains and sown towards the end of the rainy season to avoid waterlogging. Therefore, in food-deficient Ethiopia, removing constraints to crop production on Vertisols is very important (Tekalign Mamo *et al.*, 1993).

In this chapter, the experience of a research project for developing and disseminating technologies for better management of Vertisols for improving productivity is summarized.

Vertisol Technology Development and Testing in Ethiopia

In some parts of Ethiopia, particularly around Debre Berhan, farmers practise soil burning to minimize waterlogging and to improve the fertility of the soil. Small mounds are created with surface soil; dung and residual straw are put inside the mounds to burn the soil, then the burnt mounds are levelled again. In another area around Inewari, farmers construct hand-made broadbeds and furrows, principally using female and child labour, to facilitate drainage. Both soil burning and hand-made broadbed making are labour-intensive operations, and they are not technically very efficient; these traditional techniques do not exploit the full potential of the Vertisols (Tekalign Mamo *et al.*, 1993).

Animal traction is extensively used for tillage in Ethiopia but the traditional plough (*maresha*), pulled by a pair of oxen, cannot invert or shape the soil so that the land tilled with the *maresha* remains covered with water during heavy rains. In order to facilitate drainage, the Ethiopian Joint Vertisol

Project (JVP)[1] developed, through a series of design experiments, a broadbed maker (BBM) by joining two *mareshas* with a crossbar about 1.5 m long, attaching a metal wing on the outside of each *maresha*, and linking the two wings with a looping metal chain from the rear. When operated by a pair of oxen, the two *mareshas* of the BBM create two furrows on two sides of a 1.2 m bed; the chain levels the soil on the bed and covers seeds when sown or planted on the bed. During heavy rain, the furrows allow excess water from the bed to drain into a sub-field or main drain at the end of the plot. This drainage technique allows early sowing and a longer growing period, thus resulting in higher yield and less erosion. The JVP has developed a suitable agronomic package (crop varieties, planting dates, fertilizer regime and crop protection measures) to complement the BBM (Mohamed Saleem, 1995). Together it is called the 'BBM package' but henceforth BBM and BBM package will be sometimes used interchangeably for brevity.

After on-station trials, the BBM package was tested on farm from 1986 to 1989 at five Vertisol sites in the Ethiopian Highlands (Inewari, Hidi, Ginchi, Dogollo and Dejen) at altitudes ranging from 1850 to 2600 m above sea level and receiving 850–1200 mm annual rainfall. Tests were performed in collaboration with a small number of farmers selected in consultation with the local Peasant Associations, which had a dominating role in rural Ethiopia at that time. These initial tests provided opportunities to verify the technical and economic performance of the BBM package and related problems. The results showed that, with the BBM package, wheat yields averaged 1.5–3.0 t ha^{-1} in different locations compared with average yields of 0.6–0.8 t ha^{-1} of traditional wheat or teff, which the BBM package replaced. The test also led to modification of some components of the package.

From 1990 to 1995, on-farm research was continued in three (Inewari, Hidi and Ginchi) of the five sites with a particular focus on the adoption behaviour of farmers. Through the local extension office of the Ministry of Agriculture (MOA) training was given to prospective participants in the BBM package including handling, dismantling and reassembling of the BBM. Additionally, in 1993, experienced and productive farmers in Inewari were contacted to recruit new farmers and train them, to encourage farmer-to-farmer diffusion. Participants were provided with improved seeds and fertilizers on credit to be repaid after the harvest of the crop; the services of the BBM were provided free of charge. One set of BBM served six to eight farmers. The credit was provided out of a revolving fund granted by Oxfam America. A committee composed of representatives from JVP, the MOA and the Peasant Associations managed the fund. In 1995, the management of the revolving fund was

[1] A consortium comprising the Ethiopian Agricultural Research Organization (formerly Institute of Agricultural Research), Alemaya University of Agriculture, Ministry of Agriculture, International Livestock Research Institute (ex-International Livestock Centre for Africa) and International Crops Research Institute for the Semi-Arid Tropics as partners.

handed over to the Peasant Associations with local MOA staff having a supervisory role.

In 1995/96, an assessment of the economic impact of the BBM package in the three research sites showed that at all three sites improved wheat replaced traditional wheat as expected but in Ginchi improved wheat also replaced teff and pulses to some extent. Cash income increased in Inewari from Birr 269 to Birr 865, in Hidi from Birr 546 to Birr 679 and in Ginchi from Birr 684 to Birr 1221. Larger holdings had higher potential for increased cash income than smaller holdings because they could devote a larger area to the BBM package without reducing the area under other subsistence crops. Because of the low cash income of the farmers, the benefit from the BBM package, which requires purchased inputs like seeds and fertilizers, could be substantially increased by relaxing the cash constraint through credit (Weber, 1996).

Also in 1995, a survey was conducted in two of the three research villages to test if farmers were willing to buy and own the old BBM sets, consisting of two wings and a chain (farmers already had *mareshas*), rather than getting free BBM service from the project, and the price they were willing to pay. In the third village, farmers were already using the BBM sets supplied by the research project on a share basis, five to six farmers sharing each set. Willingness to buy and own indicates the farmers' confidence in the technology and interest in its continued use. The survey revealed that over 150 farmers were willing to buy the 90 BBM sets available, so the sets were sold through a lottery among interested buyers present on a pre-arranged day at each location. New BBM owners used it themselves, lent it to relatives and neighbours and, in some cases, rented it out. This was also an indication that farmers with traction animals could earn extra income by renting out BBM services to those without traction animals or with inadequate traction animals.

The Adoption Pathways for the BBM Package at Research Sites and Related Factors[2]

During the on-farm research, information on the BBM package was made accessible to all the farmers in the research villages yet it was observed that some farmers participated in the research process for different durations either continuously or discontinuously; some did not participate, some did not know how the BBM package functioned. So, from 1995 to 1996, 585 households, were surveyed from the three research villages to understand adoption behaviour and related factors. Empirical studies on agricultural technology adoption generally divide a sample into adopters and non-adopters and analyse the reasons for adoption or non-adoption at a point in time. However, a simple distinction between adopters and non-adopters was inadequate in the

[2] This section has been derived primarily from Jabbar *et al.* (1998).

present study as about half of the non-adopters had not yet acquired sufficient knowledge about the BBM package while the other half had acquired knowledge but had not yet decided to adopt.[3] Among adopters, about two-thirds used the package discontinuously and one-third continuously (Fig. 12.1, Panel A). Acquisition of knowledge and information precedes any decision to adopt so the sample was first divided into those who knew about BBM and those who did not (Panel B, Fig. 12.1). Results show that 9% of the sample did not know about the BBM package, 91% knew about the package, of which 89% adopted; among adopters the use pattern (continuous vs. discontinuous) was the same as that in Panel A. Therefore Panel A cannot be considered to correctly depict the sequence of learning and adoption. Panel B shows a more appropriate sequence: farmers move from learning to adoption to continuous or discontinuous use.

Several logistic regression equations were estimated to identify factors influencing farmers' probability of acquisition of knowledge, probability of

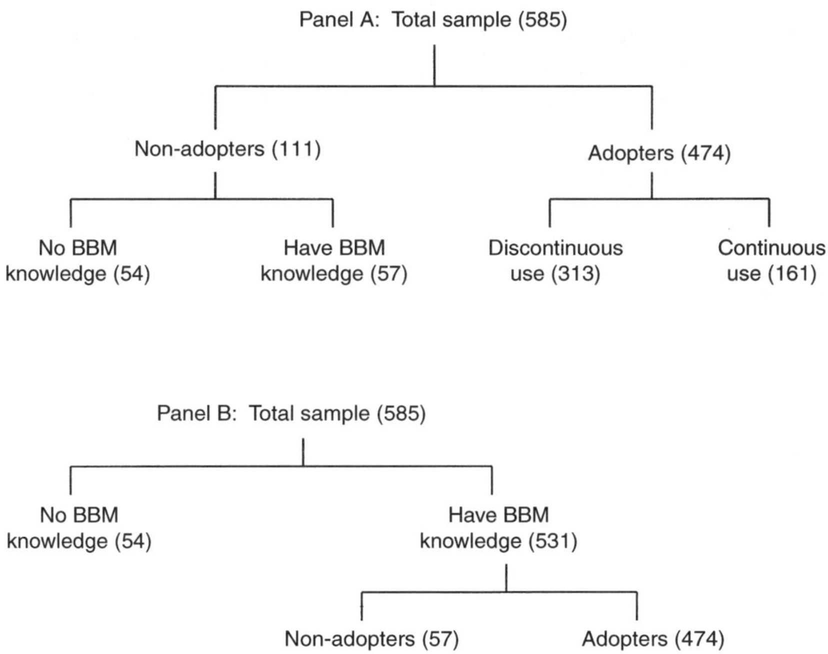

Fig. 12.1. Distribution of sample households according to knowledge, adoption and use pattern of the BBM package at three research sites.

[3] A producer is considered to *know* about a new technology if his/her acquired information reaches a threshold level. In the present case, the threshold level of information was not directly observable, so a farmer was considered to have knowledge about the BBM package if he/she heard about the BBM package and its functions and/or saw it functioning. At this stage acquisition of information was the key, acquisition of operational skill for the BBM was not yet an issue.

adoption and probability of continuous use of the BBM package on the basis of classification Panel B in Fig. 12.1. A summary of the best-fit model estimates is shown in Table 12.1. The model correctly predicted 91% of cases in terms of BBM knowledge, 92% of cases in terms of BBM adoption and 78% of cases in terms of BBM use pattern.

In general, there were significant differences among the three sites in terms of acquisition of knowledge, adoption and use patterns. Compared with Inewari, a farmer located in Hidi or Ginchi was less likely to have acquired BBM knowledge. Among those who had BBM knowledge, a farmer located in Hidi was many times more likely to have adopted BBM while a farmer in Ginchi was significantly less likely to have adopted BBM. Among adopters, a farmer located in Hidi or Ginchi was significantly less likely to have used the package continuously. The discontinuous use was more pronounced in Ginchi.

These differences might be because farmers in Inewari use handmade broadbeds, so they probably were generally more eager to learn about a better substitute and use it. Also the farmer-to-farmer training programme organized in Inewari in 1993 gave farmers there a better opportunity to learn compared with the other two locations. Inewari and Hidi farmers also had more regular access to credit compared with those in Ginchi. Some of the other factors, or their interactions, which might have influenced differences among the three sites are size of land, extent of Vertisols and waterlogging problem, animals owned and education (Table 12.2).

Table 12.1. Significant factors influencing acquisition of BBM knowledge, BBM adoption and BBM use pattern.

Factors	BBM knowledge	BBM adoption	BBM use pattern
Location	Yes	Yes	Yes
Education	–	–	+
BBM training		+	
Age of household head			
Cropland area	+	+	
Vertisol area	+		+
Waterlogged area		+	+
Family size	–	+	
Distance from market		–	
Work animals	+	–	
Perceive BBM has problem			+
Expected extra yield			
Access to credit		+	+

BBM, broadbed maker.
+ indicates positive influence on the dependent variable, – indicates negative influence, blank means non-significant.

Education, area of cropland, area of cropland under Vertisols, number of work animals, family size and distance from market had a significant influence on whether a farmer had acquired BBM knowledge or not. Household heads with better education (primary level or over) were less likely to know about the BBM package than those with no formal education. Households with a larger cropland area and area under Vertisols and a larger number of work animals were more likely to have acquired knowledge about the BBM package. Area under Vertisols had the most dramatic effect on the odds of a farmer acquiring knowledge about the BBM package: with one unit increase in the area under Vertisols, the odds of a farmer knowing about the BBM package increased 4.5 times. Since the BBM is specifically meant to address the problem of Vertisols, the high degree of influence of this variable on farmers' willingness to learn about BBM would normally be expected. The positive effect of number of work animals on acquisition of BBM knowledge might be explained by the fact that a pair of animals was required to pull the BBM, so farmers with two or more animals were perhaps more interested in knowing about the BBM than those having only one animal or no animals.

Larger family size decreased the odds of learning about BBM to some extent, perhaps because larger family labour supply decreased the need for alternative technology. Greater distance from market also decreased the odds of learning about BBM, perhaps because the transaction costs of acquiring knowledge increased with distance and also information to distant parts of the research areas might have trickled down slowly.

Among those having knowledge about the BBM package, location, education, BBM training, cropland area, area with major waterlogging problem, distance to market and work animal ownership had a significant influence on whether the BBM package was adopted or not. The odds of adoption decreased as the level of education increased while skill training in BBM increased the odds of adoption several times. Some adopters actually did not initially acquire the skill to operate the BBM, they hired somebody else to operate it. A typical example would be a farmer without BBM operational skill and another farmer with skill joining together with their *mareshas* to make the BBM.

Farmers with larger cropland areas and larger areas with major waterlogging problems were more likely to have adopted BBM. Although the area

Table 12.2. Some characteristics of the three research sites. (Source: field survey.)

Attribute	Inewari	Hidi	Ginchi
Cropland per farm (ha)	1.45	1.75	2.95
Vertisol area (%)	49	51	91
Area with major waterlogging problem (%)	19	17	42
% household heads with primary or higher-level education	59	61	38
Work animals per farm	1.66	2.21	2.17

under Vertisols significantly increased the odds of a farmer acquiring knowledge about BBM, it had no influence on adoption. Instead area with a major waterlogging problem significantly increased the odds of adoption. In the sample sites, 60% of the cropland was under Vertisols, nearly 50% of cropland had some waterlogging problem but only 23% of cropland suffered from a heavy waterlogging problem that would benefit from a BBM-type technology package.

Greater distance to market decreased the odds of adoption, perhaps because distance adds to the costs of a new technology and reduces potential net benefits. Ownership of a larger number of work animals also decreased the odds of adoption, a characteristic rather difficult to explain except that work animal ownership and cropland are highly correlated and cropland has a strong positive influence on adoption.

Among those who adopted the BBM package, area under Vertisols, area with major waterlogging, perceptions about problems with the BBM technology package and access to credit had a significant influence on whether the BBM package was used continuously or discontinuously. A higher level of education increased the odds of continuous use but BBM training had no influence on use pattern. Both area under Vertisols and area with major waterlogging increased the odds of using the BBM package continuously, which would be expected. The odds of continuous use were higher for farmers who perceived that the BBM package had some problems or disadvantages compared with those who did not perceive such difficulties. This was an apparently unexpected result but could be explained by the fact that those who used the BBM continuously and for a longer period also were more likely to have experienced or detected snags in the package. The most important problem reported by farmers was about the heaviness of the BBM unit. The other drawback mentioned by a few was the unsuitability of the BBM for saturated soil and weeds. Longer duration of access to credit for the BBM package significantly increased the odds of continuous use among adopters.

Factors Influencing Adoption Outside Research Villages

Since 1992, the government has gradually introduced market liberalization policies and a drive to achieve food self-sufficiency. Consequently a congenial environment has emerged for diffusion and adoption of improved technologies. Responding to this opportunity, the MOA and several non-government organizations, including Sasakawa Global 2000, have started diffusion of the BBM package alongside other improved technologies. A private manufacturer of BBM, who was formerly an International Livestock Research Institute technician involved in the design and testing of BBM, is also active in the diffusion effort through selling BBM sets as well as imparting training to local blacksmiths in the fabrication of the equipment. This manufacturer alone claimed to have sold over 12,000 sets of BBM by 1997 and participated with other agencies in training 26,000 farmers in using the BBM[4] (Chapotchka, 1997). Including supplies from government workshops, about 15,000 BBM sets have been reportedly distributed by different agencies to different parts of the highlands of Ethiopia and Eritrea until 1998. Moreover, there are reports that BBM has been adapted in several lowland areas in the country for ridging to conserve water rather than for draining excess water. However the number of BBMs actually in use and the number of farmers using them in each year are yet to be ascertained.

In order to understand factors influencing adoption of BBM in the wider economy, a survey was conducted among adopters and non-adopters in two areas in 1996/97. Using probit and tobit regressions on the survey data, it was found that access to credit, history of fertilizer use and high school education significantly increased probability and intensity of adoption; and number of draught oxen ownership, distance from market and area of teff production significantly decreased the probability and intensity of adoption. The intensity of adoption was measured by the area devoted to the BBM package (Gezahegn Ayele and Heidhues, 1998). These results are fairly consistent with the findings presented above from research sites except that, in this study, only adoption and non-adoption were the focus of analysis, knowledge acquisition was not.

Externality and the Need to Move from Plot to Watershed

During on-farm testing, it was observed that, although the BBM solves the problem of waterlogging of the plot on which it is used, sometimes it also creates increased waterlogging on the plot downstream. This is described as the problem of negative externality of a technology. This problem is more serious on lower slopes, where waterlogging is generally worse than on higher slopes, where water drained from a plot is likely to move quickly further down the slope. This may create conflict among farmers and may work as a disincentive for adoption. If a BBM is used on a number of contiguous or scattered plots in a landscape or large field, finding or creating appropriate common outlet(s) for drained excess water from individual plots can solve the problem of externality. This may require creation of field drains as well as longer common drains into which field drains will carry excess water. Implementation of such a strategy requires land and water use planning at micro-watershed level through participation of the community in design, development and maintenance of common goods, e.g. drains. If the BBM plots are fairly close or contiguous, creation of common outlets may be easier than if they are scattered.

---

[4] A separate report quoted the same supplier as selling at least 25,000 sets of BBM (Schioler, 1998). Perhaps this higher figure refers to the number of farmers using BBM, which is consistent with the number of farmers trained, rather than the number of sets sold. Normally a set is shared by a number of farmers.

Experience on Watershed Management

In 1994, the JVP started work on a pilot watershed in Ginchi Wereda in the highlands. The objectives of this project were, *inter alia*, to assess if negative externality in BBM use can be overcome by creation of common drains through community participation, to assess how the community can organize itself to undertake such tasks voluntarily and to assess which factors may motivate farmers to participate and contribute in the creation of such a local common good. The project provided the opportunity to test standard theories of collective action in local resource management.

The pilot watershed comprises about 350 ha, which itself is part of a Peasant Association (PA) of about 250 households with about 800 ha of land area. Within the 350 ha watershed, a sub-watershed of about 50 ha of mostly Vertisols, held by a total of 57 households, was delimited by the JVP to construct a main common drain with the participation of landowners concerned (the beneficiary group). The possibility of extending the drain to the whole watershed area in subsequent years was explicitly envisaged if the experience was to prove a success.

The construction of the main drainage channel clearly involved important indivisibilities. Unless a critical amount of labour input was allotted to this task, the channel would not be completed or yield any benefit in the form of increased land productivity. On the other hand, no additional labour contribution beyond that critical amount could give rise to productivity increments. Since landholdings are widely dispersed across the sub-watershed, there was no possible sub-coalition of farmers with lands concentrated in a well-delimited part of this area. As a result, it was not conceivable that only a portion of the drainage system would be constructed to the benefit of a fraction of the farmers and at the expense of other farmers whose lands were located below the drain. Either the channel was to be constructed across the entire sub-watershed area with the required technical specifications, or it was not to be produced at all. To put it in another way, due to scattered ownership, the channel within the sub-watershed area could not be divided into several segments that could be the object of separate construction decisions.

The proposed central drainage channel was built during May and June 1995 by labour-intensive methods. There was a good measure of uncertainty regarding the exact amount of labour input required and the timing of the efforts involved. Farmers may well have believed that their contributions were likely to be sufficient to complete the drain within the prescribed time given their initial expectations concerning the advancement of the work and the advent of the rains. But, when the digging was started, it soon became clear that the soil was very hard for manual work. Progress was therefore much slower than expected. In consultation with the farmers and only after it was clear that the problem was real and serious enough to prevent completion of the drain during the season, the project's staff decided to use a tractor to loosen the soil and the farmers then removed the earth and shaped the drain. During the first season after completion of the drain, BBM plots were more effectively drained and gave higher wheat yields; teff plots within the micro-watershed could be drained better and gave higher yields.

Determinants of Individual Household Contributions to Drainage Construction[5]

In order to identify important factors influencing participation, tobit regression was used on data collected from 53 of the 57 households in the watershed. The dependent variable was participation, measured by the amount of time spent by each household in earth works as well as in attending preparatory meetings in which operational procedures and rules for the drainage project were discussed and decided. The results of the best-fit model indicate that the intensity of participation by the sample households in the construction of the drainage system was related significantly to the following factors. First, the amount of land owned in the watershed. The more land that is owned the greater the incentive is to contribute to an investment that is designed to increase land productivity.

Second, the relative number of parcels owned that form a compact set with one border along the channel. The location of the parcels can be thought to bear upon the benefits derived from the drainage infrastructure in two different ways. On the one hand, the gradient of the parcels may matter since drainage intervention is likely to be needed less and less as lands are located higher up the watershed's foothills. On the other hand, proximity of the parcels to the drain may be expected to bring large benefits if water is drained away more effectively close to the central channel. Moreover, it can be presumed that, when a farmer owns a large compact landholding formed by a set of contiguous parcels having a border along the drainage channel, he is able to organize, at lower costs, a system of secondary drains connected to these channels.

Third, the potential effectiveness of cultivation measured by the capital/labour intensity of owned factors. Households with a higher capital : labour ratio showed a higher tendency to contribute to the public good creation because of the possibility of deriving higher benefits from the improved technology, which also requires higher capital input.

Fourth, leadership quality. The positive effect of leadership on participation indicates that knowledge of the potential gains from collective action may be more easily forthcoming when farmers are members of the executive committee of the local PA. Farmers who are more aware of potential gains from cooperation in general tend to assume responsibilities and leadership roles in PAs. It is noteworthy that members of the executive committee of the

[5] This section has been derived primarily from Gaspart *et al.* (1998).

local PA have contributed an average of 22 h, compared with 12 h for all participating households, and with 7.5 h for all sample households.

In short, what appears most important is that, with the exception of the leadership factor, individual household labour contributions to the public good are well explained by factors that clearly determine the potential personal benefits that different households can expect to draw from the creation of the drainage infrastructure. These results are quite close to those arrived at by White and Runge (1995) in a similar watershed management context. They found that cooperation among beneficiaries emerged in many cases and, within a given watershed, practical knowledge of the potential gains from collective action (in this case, the building of a soil conservation infrastructure to prevent externalities in the form of water spillover) and membership in indigenous peasant organizations are the best predictors of landholder choice to cooperate (White and Runge, 1995, pp. 1697–1698). In fact, a careful examination reveals that it is not only awareness of potential benefits but also the ability to draw those benefits that motivates landholders to cooperate.

Current Research: from Vertisols Management to Integrated Resource Management

Apart from the BBM package, the JVP also conducted on-station and on-farm research on technological solutions to other problems, e.g. animal feed, nutrition, soil fertility management, etc. However, these component technologies have been tested mainly at the scale of plot, animal or farm, often separately. Multiple interventions and their integrated effects were not assessed. The highland ecosystems, like any other ecosystem, have been viewed primarily as biophysical systems with geochemical and biological functions or, at best, as human production systems with product yields or economic returns as the focus. In reality, farmers have many enterprises and, in any year, they may cultivate many small plots that are spread across a landscape. Sharp and sudden changes in landforms within short distances (steep slopes, moderate slopes, flatlands and valleys) are very specific to the highlands. Rugged terrain limits accessibility, and soils along higher slopes are normally shallow. A combination of landforms and soil types within short distances creates 'niches' – land potential and land-use practices whose constraints are different when the gradient and altitude are combined to divide a landscape (Fig. 12.2). Consequently, potential solutions will also be different for different niches.

However, changes occurring in one part or 'niche' of a highland ecosystem have impact on other parts. Individuals and communities decide to use different parts of a landscape in particular ways and such use may be complementary or competitive. They also have environmental implications both on site and beyond, for the short as well as the long term. For example, technology interventions in the Vertisol-dominated part of an ecosystem may generate changes in the pattern of use in other parts of the landscape. Therefore,

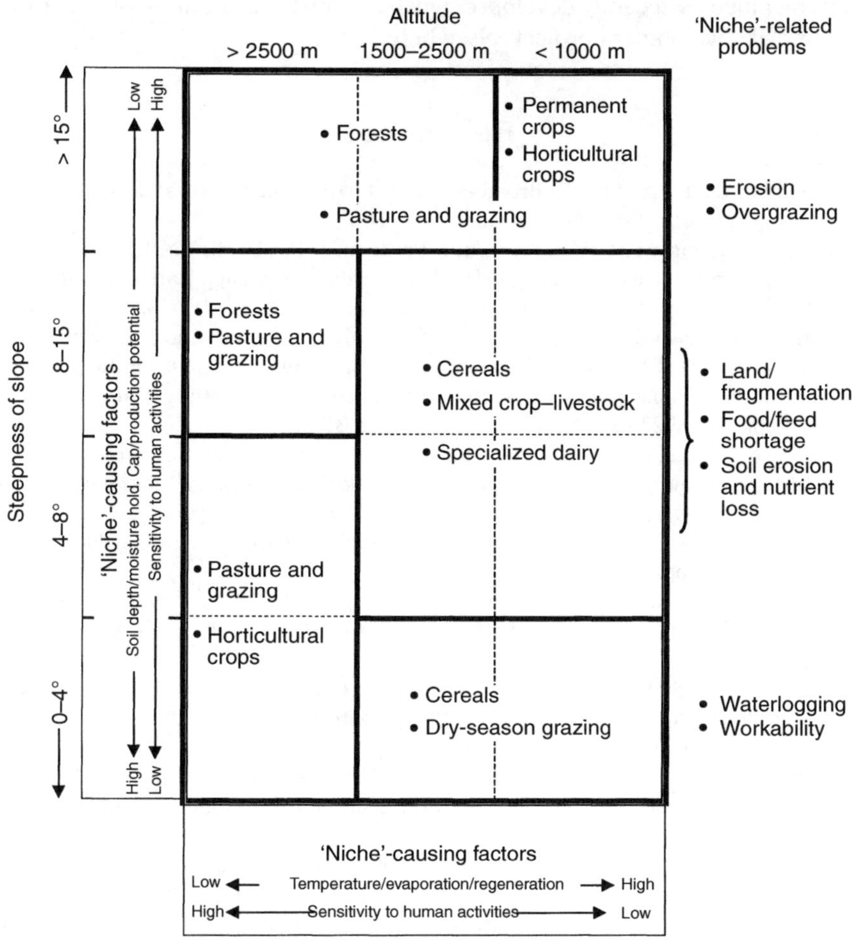

Fig. 12.2. Land-use practices and major constraints for production improvements in the highlands.

interrelationships between biophysical and human dimensions need to be integrated both spatially and temporally to identify ways to improve conditions of the ecosystems and human welfare. Bioeconomic modelling is a useful tool for integrating the biophysical and socio-economic aspects of a landscape, e.g. a watershed, and for assessing the multidimensional consequences of several technology and policy interventions.

Recently, the JVP has expanded its watershed research agenda to include more holistic and farmer participatory assessment of the consequences of multiple technology interventions as well as development of a bioeconomic model of the watershed to assess, in an integrated manner, the long-term consequences of different technology and policy interventions. It is expected that the model, once fully developed, will be useful for evaluation of integrated watershed development projects elsewhere.

References

Chapotchka, Ekaterina (1997) Broad beds and furrows: agricultural development in Ethiopia. *Culture Crossroads* October, 177–185.

Gaspart, F., Jabbar, M.A., Melard, C. and Platteau, J.-P. (1998) Participation in the construction of a local public good with indivisibilities: an application to watershed development in Ethiopia. *Journal of African Economies (Oxford)* 7(2), 157–184.

Gezahegen Ayele and Heidhues, F. (1998) Analysis of innovation, dissemination and adoption of Vertisol technology: some empirical evidence from the highland of Ethiopia. Paper presented at the Conference on Soil Fertility Management in Ethiopia, April 21–23, Addis Ababa, Ethiopia. 37 pp.

Jabbar, M.A., Hailu Beyene, Mohamed Saleem, M.A. and Solomon Gebreselasie (1998) *Adoption Pathways for New Agricultural Technologies: an Approach and an Application to Vertisol Management Technology in Ethiopia.* Socioeconomic and Policy Research Working Paper No. 23. International Livestock Research Institute, Addis Ababa, Ethiopia, 27 pp.

Mohamed Saleem, M.A. (1995) Fragile East African highlands: a development vision for smallholder farmers in the Ethiopian highlands. *Outlook on Agriculture* 24(2), 111–116.

Schioler, E. (1998) Down-to-earth research. In: *Good News from Africa – Farmers, Agricultural Research and Food in Plenty.* International Food Policy Research Institute, Washington, DC, USA, pp. 38–41.

Sutcliffe, J.P. (1993) *Economic Assessment of Land Degradation in the Ethiopian Highlands: a Case Study.* Ministry of Planning and Economic Development, Addis Ababa, Ethiopia.

Tekalign Mamo, Abiye Astatke, Srivastava, K.L. and Asgelil Dibabe (eds) (1993) *Improved Management of Vertisols for Sustainable Crop–Livestock Production in the Ethiopian Highlands.* Synthesis Report 1986–1992. Technical Committee of the Joint Vertisol Project, Addis Ababa, Ethiopia, 199 pp.

Weber, M. (1996) The potential impact of the broad bed maker in Ethiopian agriculture: a cross-site comparison over three study sites in the highlands. Masters thesis, Humboldt University, Berlin, Germany, 124 pp.

White, T.A. and Runge, C.F. (1995) The emergence and evolution of collective action: lessons from watershed management in Haiti. *World Development* 23(10), 1683–1698.

Part III

International Perspectives on the Management of Vertisols

Low-cost Animal-drawn Implements for Vertisol Management and Strategies for Land-use Intensification

13

A. ASTATKE AND M.A. JABBAR

International Livestock Research Institute, Addis Ababa, Ethiopia

Key words: East Africa, Ethiopia, Vertisols, animal power, implements, soil management, land use

Introduction

From an estimated 300 million ha of Vertisols found globally, 43 million ha are located in tropical Africa. An additional 80 million ha of soils with vertic properties require similar management to Vertisols to exploit their agricultural potential (Virmani, 1988).

Most of Africa is characterized by marginal rainfall but the highland areas of East Africa with altitudes above 1500 m above sea level provide cooler temperatures and high rainfall. Where soils are deep, the growing period is more than 180 days and they are suitable for a variety of crops and livestock (Morgan, 1972). Ethiopia has 50 million ha of land in the highlands representing about 62% of the East African highlands. Of this area, 16% is estimated to be Vertisols, and about an equal amount has vertic properties. These soils are generally found on the lower landscape on slopes not exceeding 8%, making them attractive agriculturally. However, there is an inherent management problem due to waterlogging caused by the slow percolation rates of these soils coupled with low evaporation. Erosion caused by runoff from the higher landscape and from draining off the excess water can exacerbate soil erosion, especially gully erosion, which is another major problem in the highlands.

(eds J.K. Syers, F.W.T. Penning de Vries and P. Nyamudeza)

Traditional Management of Vertisols and Land-use Types

During the main growing period, the severity of waterlogging varies from area to area depending on the clay content of the soil, the amount of rainfall and the soil temperature (which also depends on the moisture content). Farmers in the Vertisol areas realize the adverse effects of waterlogging on crop productivity and have developed traditional methods to overcome them.

A traditional method to overcome waterlogging is planting crops late in the season to grow on residual moisture, after the excess water has drained away naturally. Early-maturing varieties of crops such as wheat, chickpea, roughpea, etc. have a short growing period of not more than 3 months. *Eragrostis tef*, which tolerates mild waterlogging, is planted during the middle of the rainy season. The traditional practice does not fully exploit the growing period and enhances erosion due to the lack of ground cover during the rainy season. Crop yields in this traditional practice are low, averaging 0.8 t ha^{-1} (Debele, 1985). In the last 10 years, the growth of highland rice, on Vertisols around Lake Tana in the northwestern part of the country, has expanded, and is said to give better yields with the traditional practice than other indigenous crops grown in the area.

In the high-altitude areas of Ethiopia, i.e. above 2400 m above sea level, a unique practice called *guie* has been adopted for growing barley on Vertisols after leaving the area fallow for 5–8 years (Tesema and Yirgou, 1973). The farmers plough the land three to four times during the dry season, heap the soil at irregular spacing and burn it with dry manure, grass and weeds. The soil is then spread back on the fields (Debele, 1985). After the onset of the following rains in mid-June, the fields are ploughed again and barley is grown. The planting of barley continues for two to three seasons and the land is then left fallow again. The burning of the soil changes the topsoil structure, producing a coarser texture, which facilitates better water movement and drainage and increases available phosphorus (Tesema and Yirgou, 1973).

Different cultivation techniques are also practised using the *maresha*, the Ethiopian plough, to minimize the waterlogging problem on Vertisols. Flat seedbed preparation is common on gentle slopes except for the fact that outside ditches are sometimes dug to control flooding. This method is common in drier regions and crops such as faba bean, field peas, barley, linseed and sorghum are planted (Tedla and Mohamed Saleem, 1992).

In some parts of the Central Highlands of Ethiopia, drainage furrows are made with the *maresha* on the flat seedbed after seeding. These furrows are made across the contour at distances ranging from 3 to 7 m. These drainage furrows have an average 20 cm width and 15 cm depth. The area taken by the drainage furrows from the crop areas can be 10–15% (Westpal, 1975). In high-rainfall areas, it is common to create ridges and furrows using the *maresha* at intervals of 40–60 cm. In this traditional ridge and furrow system, the furrows take up 40–50% of the crop area.

At Inewari, in the Central Highlands of Ethiopia, surface drainage of Vertisols is facilitated with the use of manually formed broadbeds and furrows. The seedbed is made by making three to four passes with the *maresha*. In the middle of the rainy season, the seeds are broadcast and furrows are made with the *maresha* at an interval of 0.8–1 m. Using family labour, the soil is then scooped up from the furrows and dumped on the beds. By using this method they not only form the broadbed and furrow but also cover the seeds. Grass drainage channels are constructed also to carry the water from the crop fields. This practice of constructing broadbeds and furrows manually requires hard work by the family. In the last 2 years, this method of manually constructing broadbeds and furrows has been extended to neighbouring Vertisol areas that previously used the ridge and furrow system.

The *Maresha*

The *maresha*, pulled by a pair of oxen, has been used in Ethiopia for a long time. There are certain pockets of the country where hoe cultivation is still practised but, by and large, cultivation is carried out by a pair of oxen pulling the *maresha*. In some areas of the country, horses and mules occasionally pull the *maresha*, but generally oxen provide the main traction force.

The *maresha* consists of a metal point or tine, which in turn is fastened to a wooden neck yoke as shown in Fig. 13.1. On either side of the metal point are two wooden wings that push the soil aside. The complete traditional plough is a light implement, ranging from 17 to 26 kg with the yoke (Goe, 1987), which

Fig. 13.1. Ethiopian *maresha* plough.

means that it can be transported to and from the field over difficult terrain by one person. Except for the metal tine, which the farmer has to buy from the blacksmith, the rest is home-made. Depending on the crop types, three to four ploughings are required with the *maresha* before a field is ready for planting.

Each cultivation pass is made at right angles to the previous one to disturb the whole soil. The depth of the first ploughing ranges from 5 to 8 cm while at the last pass up to 20 cm can be attained. For land preparation and seed covering with the *maresha*, on average 100 h ha^{-1} and 150 h ha^{-1} are required for Vertisols and light soils, respectively (Astatke and Mathews, 1982). The power from a pair of local zebu oxen, each weighing not more than 300 kg, ranges between 0.50 and 0.90 kW (Astatke and Mathews, 1982).

A serious disadvantage of the *maresha* is its inability to cover seeds adequately. In the traditional cultivation method, seeds of all the cereal crops and pulses are broadcast before the last pass with the *maresha*, which is then used to cover the seeds. The exception is *Eragrostis tef*, which is broadcast and left. Thus, the depth of the coverage varies from seeds not covered at all to the maximum depth that the *maresha* tine penetrates. This might be the reason why farmers tend to use high seed rates as germination rates can be low. In some pockets of the central highlands, it is common to use a wheat seed rate of 250 kg ha^{-1}, which could be 15–25% of the expected total production.

The Broadbed Maker (BBM) and its Evolution

The International Crops Research Institute for the Semi-Arid Tropics (ICRISAT), based in India, developed an animal-drawn implement in the mid-1970s for making broadbeds and furrows to improve the surface drainage of Vertisols. This implement proved functionally effective but it required more power than a pair of zebu oxen could generate. It was also expensive, more expensive than the Indian subsistence farmers could afford; therefore adoption was very poor. When the equipment was tested in Ethiopia, farmers had the same opinion.

In 1986, the International Livestock Centre for Africa, now the International Livestock Research Institute, the Ethiopian Agricultural Research Organization (previously the Institute of Agricultural Research), Alemaya University of Agriculture and ICRISAT formed a consortium, called the Joint Vertisol Project (JVP), to develop low-cost technologies to improve the management and productivity of Vertisols. Ethiopian highland farmers have limited cash income, so any improvement in tools and farming methods must take this into consideration and make economy a priority. Implements and technologies designed, therefore, must have minimum cost so farmers can afford them.

From the beginning, researchers consulted farmers on design. Tests conducted on farm and farmers' suggestions derived from such tests have been invaluable for refining the implement. As researchers took the implement beyond the areas where it was tried on farm, they realized they needed to refine it further to tailor it to specific farmers' circumstances, which included topography of their plots, length differences between the *maresha* beams, etc.

The first land-shaping implement that the project developed had a wooden wing that functioned as a mouldboard, replacing the traditional flat wings of the *maresha*. After loosening the soil (which took three or four passes with the *maresha*), the implement lifted and heaped the soil to form a bed. However, it was no faster than the *maresha* in making the raised beds, and an untrained person using the new implement found it difficult to make a seedbed of good quality. Hence, farmers rejected this modification of the *maresha* as well.

The second version was a BBM made from the *maresha* (Fig. 13.2). The main beams of the *maresha* were shortened to about 90 cm and were connected with a simple wooden frame. Mouldboard-shaped wings, two bigger ones scooping the soil to the centre and two smaller wings expelling it away, forming beds and furrows, replaced the two flat *maresha wings*. This implement could be used only for shaping the land, and therefore the field had first to be ploughed three or four times with the *maresha*. The implement weighed about 35 kg, depending on the type of wood used. The average power required to pull it was 0.7 kW – the same as the power required to do the first ploughing with the maresha (Jutzi *et al.*, 1986). However, when farmers tried it out in the field, they found it too heavy to handle and too bulky to carry to and from the field. Also, finding a spanner and 12 bolts to construct the wooden BBM frame was not easy in the rural areas. So researchers went back to the drawing board to redesign the implement.

Fig. 13.2. Second version of the broadbed maker.

The present version of the BBM is recommended widely, the implement having proved effective and acceptable in several years of on-farm trials. It is based on the traditional wooden frame plough, to which is attached the *maresha* metal plough-share with its land-shaping mouldboard. Two *mareshas* are connected in a triangular structure, mounted about 120 cm apart on a crossbar (Fig. 13.3). The top ends of the *maresha* beams are tied together and connected to the yoke as in the traditional method. To maintain the distance of 1.2 m between the *maresha* tips, a crossbeam is tied between the two *maresha* poles a metre from their lower edges. A steel mouldboard wing is attached on each of the inner flat wings.

As a pair of animals pulls the BBM through the soil, it makes two parallel cuts. The wings scoop the soil towards the middle and mound it, forming the broadbed and furrow. The chain attached at the edge of the metal wings and which connects the wings not only shapes the beds evenly, but also covers the sown seeds. Seeds are broadcast as early as June. The result is a series of broad, raised and seeded beds 15 cm high by 80 cm wide, separated by 40-cm-wide furrows.

The power requirement of 0.62 kW for this new BBM is lower than for the wooden BBM, as the frictional force of the metal wings passing through the soil is less than that of the wooden wings. In 6 working hours, the area shaped into broadbeds and furrows with this BBM pulled by a pair of oxen was 1.2 ha compared with only 0.4 ha when a wooden BBM was used.

The potential effect of this tilling device on Vertisols and their food-producing capacity is immense. Benefits will be long-lasting. Early planting helps to establish a vegetative cover early in the season and thus reduces soil erosion. In Vertisol areas, the BBM technology package has doubled crop

Fig. 13.3. The latest version of the broadbed maker.

yields tripled fodder yields and increased the productivity of farm labour by 50–75% (Asamenew *et al.*, 1988). In the 1995 cropping season, wheat grain yields from the 50 farmers that used the BBM and its package on 0.5 ha each at Gimbichu, in the Central Highlands of Ethiopia, more than doubled their grain yields compared with the traditional average yield of 1 t ha^{-1}, as shown in Table 13.1. Replacing local wheat with the ET-13 wheat variety on broadbed and furrow (BBF) land systems improved grain energy output from 6.3 GJ ha^{-1} to 13.2 GJ ha^{-1} (Mohamed Saleem and Astatke, 1996).

At present several thousand farmers are using BBMs to grow wheat on highland Vertisols. Adoption pathways followed by farmers, particularly in the on-farm research sites, are described elsewhere. There are several concerns about the BBM that are being raised by some farmers and extension organizations, which are related mainly to lack of training. Some farmers have used the BBM with their traditional late planting method, requiring higher amounts of power and affecting the quality of the beds constructed. Late distribution of fertilizer in the middle of the main rainy season by private and governmental agencies was not a problem for the traditional cropping system but has become a hindrance for BBM adoption. The period for using a BBM is short, optimum dates ranging from 10 to 15 days after the start of the main rainy season, thus limiting its use. In some instances, the field drains constructed by the BBM are taken as the main drainage network without the construction of primary and secondary channels for proper diversions for controlling runoff, affecting the evacuation of the excess water from the fields.

Early planting with a BBM is not without its problems. As few farmers adopt at the beginning, more pests are harboured by a few BBM plots in a field, especially if there are dry spells. In some cases, better field drainage by the BBF is said to have caused more weed infestation, which would mean more labour input for weeding. But by and large farmers who are well trained on the BBM package have expressed their satisfaction with the results obtained.

Additional Attachments to the BBM for Improving Tillage Efficiency and Resource Conservation

Early planting resulting in higher yields than traditional late-planted crops is the main advantage of the BBM package. It has other advantages as well, e.g.

Table 13.1. On-farm wheat grain yields (Mg ha^{-1}) at Gimbichu using the BBM package in the 1995 cropping season. (Source: ILRI, 1996.)

No. farmers	Range	Mean
3	2.5–3.0	2.9
10	3.0–3.5	3.4
20	3.5–3.8	3.7
17	3.8–4.6	4.3

the BBM crop is harvested early during the severe food deficit period, so it contributes to food security. The price may be higher at this early stage so higher cash income from any sale is another advantage. However, the JVP diversified its research effort to make the equipment more versatile, so that its technical and economic efficiency might be improved further.

Currently, the broadbeds are ploughed and reconstituted if the BBM package is to be used in the next season. The possibility of retaining the broadbed for repeated use with minimum tillage was considered. This would save animal and human labour for various tillage operations. Concomitantly, the possibility of line seeding rather than broadcasting was considered because the traditional method of broadcasting seeds and covering with a *maresha* has been shown to mix 15.3% of broadcast wheat seed to a depth of 10–20 cm, yet leaving 25.3% within the top 2.5 cm (Tinker, 1989). Due to this variation of coverage depth the germination rate is low, which is the main reason why farmers use high seed rates. The BBM is better than the traditional plough in covering seeds but more efficiency can be gained by line sowing. This could reduce the required seed rate by improving the germination rate and also reduce the required fertilizer rate by improving nutrient intake by plants. Further advantages could be better control of weeds (making weeding easier and less labour-demanding) and stubble incorporation into the soil, thereby partially filling the cracks, thus reducing moisture loss and helping the following crop.

Several options to achieve the above goals were considered and modifications of the BBM with additional attachment(s) have been designed and tested. First, a version that has proved technically efficient in on-station trials over 2 years aimed at minimum tillage is a blade harrow consisting of a metal blade 4 mm thick fixed on both sides of the *maresha* tines. This blade harrow uniformly cuts the soil on the BBF at about 5–8 cm below the surface, thus slicing weeds at the rooting level when the soil condition is moist, during the small rains or at the beginning of the main rains. This is used for postharvest cultivation, then the broadbeds are left until the early rains for planting with minimum soil disturbance. At this period, the power requirement for the implement is reduced drastically. Time for land preparation and reconstruction of the BBF is not required every year if minimum tillage is used.

Second, a planter attachment to the BBM has been developed for line seeding/planting. This is similar to the traditional planter used by the Afars in the eastern part of Ethiopia; it is a simple hand-metered seeder, made of calabash and bamboo, which is mounted on the *maresha*. It has been redesigned to be attached to the BBM with a sheet metal funnel and four connecting tubes. A set of plain tines and tines of leading and trailing coulter units at a 45° penetrating angle to the horizontal soil surface is used. The funnel consists of a circular hopper 100 mm in diameter with a disked bottom drilled with four equally spaced 25 mm diameter holes to which coulter tubes are attached. A centre tube supports four baffle plates with a 70 mm diameter shallow cone placed above, which has now been changed to a double-layered cone as it provides better uniformity in seed distribution. The holes inside the hopper can be blocked off according to the new arrangement. The seeder is supported by a bar clamped to the tine bar.

The blade harrow and the planter can be used either together or separately depending on the choice of the user. Trials conducted in 1997 and 1998 showed that the time required for seedbed preparation with the minimum tillage system was 15 h ha^{-1}, a quarter of that required by the traditional system for the production of wheat plots. In 1997 trials, height at harvest, straw and grain yields of durum wheat were significantly higher with the minimum tillage practice than the conventional or the traditional systems even though vigorous growth led to higher plant lodging (Table 13.2) and production of smaller seeds – 33.2 g and 35.7 g per 1000 seeds for minimum and conventional tillage systems, respectively. The minimum tillage system still produced significantly higher wheat grain yields than both the conventional and traditional systems. The 1998 wheat grain yield was affected and lower than normal yield due to unexpected rains during the flowering stage of the wheat growth at Debre Zeit; the mean grain and straw yields under minimum tillage were still significantly higher than the conventional tillage system (Table 13.3). Inputs of seed and fertilizer rates were reduced in both years on average by 25 and 30%, respectively, using the funnel seeder.

Table 13.2. Effect of different tillage systems on height at harvest, grain and straw yields (mg ha^{-1}) of durum wheat variety at Debre Zeit in 1997. (Source: EARO, 1998.)

Tillage systems	Planting methods	Lodging (%)	Height at harvest (cm)	Grain yield	Straw yield
Minimum	Funnel planter*	63	129^{a}	1.94^{a}	4.14^{a}
Conventional BBM	Funnel planter	33	117^{b}	1.67^{b}	3.74^{b}
Traditional	Broadcast	5	120^{b}	1.72^{b}	3.63^{c}

*Seed and fertilizer mixed.
Values followed by same letters in a column are not significantly different ($P < 0.05$).

Table 13.3. Effect of different tillage systems on height at harvest, grain and straw yields (mg ha^{-1}) of durum wheat variety at Debre Zeit in 1998. (Source: EARO, 1999.)

Tillage systems	Planting methods	Height at harvest (cm)	Grain yield	Straw yield
Minimum	Funnel planter*	106.6^{ab}	1.35^{a}	3.09^{ab}
Conventional BBM	Funnel planter	100.4^{bc}	1.01^{b}	2.61^{bc}
Traditional	Broadcast	95.8^{c}	0.90^{b}	2.34^{c}

*Seed and fertilizer mixed.
Values followed by same letters in a column are not significantly different ($P < 0.05$).

In a separate on-station trial, zero tillage (with herbicide) was compared with traditional tillage on both flat and broadbed plots. Results showed differences between tillage methods, so zero tillage saves all the expenses related to seedbed preparation. These options will be tested on farm for 2–3 years beginning from the 1999 cropping season. The test will be conducted at a site where farmers have adopted the BBM package and so are familiar with the operational mechanism and advantages and disadvantages of the package. In conducting the test and selecting volunteers for participation in the on-farm test, based on theories and practices in adoption of technologies, it is hypothesized that:

1. Although, on station, zero tillage performed similarly to traditional tillage with labour-saving benefits, farmers may not like to test it alongside minimum tillage. There is a need to test these options separately and in a stepwise manner.
2. Farmers who adopted the BBM package are more likely to volunteer for testing the minimum tillage equipment and/or the planter than farmers who are yet to adopt the BBM package, because adopters confident about the benefits of the BBM package may like to test the possibility of enhancing that gain.
3. Farmers, with or without BBM experience, would be reluctant to test zero tillage at the beginning as zero tillage is not consistent with traditional practice.
4. If the on-farm tests with minimum tillage and/or planter show promising results, farmers are more likely to volunteer for testing zero tillage as a logical sequence. Farmers experiencing good results from participating in minimum tillage and/or planter trials are more likely to volunteer for zero tillage tests than those who are yet to test minimum tillage and/or the planter.

Consultation meetings were held with farmers in a selected research village to share the findings of the on-station trials on zero and minimum tillage and the planter attachments to BBM. Then volunteers were sought to participate in one or more of the trial components. Out of 15 farmers participating in these discussion meetings, 12 wanted to know more about these trials and to see the equipment and the way it functioned. These farmers were then invited to the research station, where the equipment and its mechanisms were explained and demonstrated, though the trial plots could not be shown as this was off season. Eventually all 12 farmers wanted to participate in the on-farm trial with minimum tillage and the planter but none was willing to include zero tillage in the trial. In case of loss of crop or failure of the experiment, trial farmers were to be compensated but even compensation was not adequate for motivating farmers to test zero tillage. Thus, so far, farmer behaviour appears to be consistent with the hypothesis postulated earlier.

Land-use Intensification Opportunities

It is projected that smallholder farming conditions will worsen as per capita land holding is expected to decline from 1.76 ha in 1985 to 1.11 and 0.66 in

2000 and 2015, respectively (IUCN, 1990), due to rapid population growth. Hence, high-potential and more resilient land needs to be used more intensively to meet human and livestock needs.

Vertisols and soils with vertic properties, which are potentially productive, cover the moderate slopes and flat areas of the highland landscape. None the less, farmers in the Ethiopian highlands underuse the Vertisols, using them for only part of the season for growing low-yielding cereals such as *Eragrostis tef*, local wheat and a few pulses. Also farmers are able to grow only a single crop of tef or pulses in a year on Vertisol plots because fields become waterlogged early in the season. Even in areas where farmers make their broadbeds by hand, they do not plant early, because the soil is too hard for them to work. The BBM can help intensify the use of Vertisols in three ways: First, early planting of high-yielding wheat or teff allows use of a longer growing period giving higher yields. This has already been demonstrated successfully and farmers are rapidly adopting this option. Second, experimentally, it has been shown that wheat can be intercropped with forage legumes, which improve the quality of the harvested fodder for livestock, without diminishing the yields of the food grain. Combined with the use of cross-bred dairy cows, the economic benefits can be enhanced further (Kassi *et al.*, 1999). Using the BBM also provides the opportunity for better use of draught oxen. Third, the early planting of short-duration crops could allow an early harvest, making a second crop possible on the residual moisture (Astatke and Mohamed Saleem, 1998). If water can be conserved in ponds or reservoirs, a sequential crop would be feasible in the same cropping season through minimal supplementary irrigation at planting time to secure germination. The same furrows of the BBFs that evacuated the excess water during the rainy season would be used as furrows for irrigation. This could be further intensified and natural resources sustained using minimum tillage and seed/fertilizer row-seeding devices.

The traditional practice of ploughing early and planting late exacerbates wide-scale soil erosion. But early planting helps to establish good ground cover, which minimizes the direct impact of rain on the soil and, compared with late-planted plots, reduces soil erosion by more than 100%. If food and fodder production per unit of land area can be increased by combining better drainage with sequential or rotational cropping of legumes during most of the year, then people should be able to meet their basic needs for food and fodder. This farming system can also bring about a positive effect on the entire landscape and the environment. The upper, steeper slopes can be spared from overuse, reversing the ultimate negative impact that overuse makes on land and biodiversity. Both the farm family and the environment benefit.

Research Needs and Opportunities

Research is needed not only for developing appropriate technologies and packages for land-use intensification but also for assessing adoption pathways. In order to assist extension and development agencies to promote the technology package, characteristics of potential adopters and non-adopters should be identified while assessing the technical and economic performance at the on-farm research stage. Active farmer participation in on-farm testing with the package is expected to reduce the time lag between technology generation and adoption because the advantages of the package would be appreciated by the participants, while pitfalls and deficiencies would also be identified and resolved in good time.

References

Asamenew, G., Jutzi, S.C., Tedla, A. and McIntire, J. (1988) Economic evaluation of improved Vertisol drainage for food production in the Ethiopia highlands. In: Jutzi, S.C., Haque, I., McIntire, S. and Stares, J. (eds) *Management of Vertisols in Sub-Saharan Africa*. ILCA, Addis Ababa, Ethiopia, pp. 263–283.

Astatke, A. and Mathews, M.D. (1982) Progress report of the cultivation trials at Debre Zeit and Debre Berhan. ILCA, Addis Ababa. (Mimeo.)

Astatke, A. and Mohamed Saleem, M.A. (1998) Effect of different cropping options on plant-available water of surface-drained Vertisols in the Ethiopian highlands. *Agricultural Water Management* 36, 111–120.

Debele, B. (1985) The Vertisols of Ethiopia: their properties, classification and management. In: *Proceedings of the Fifth Meeting of the Eastern African Sub-Committee on Soil Correlation and Land Evaluation* (Sudan, 1983). World Soil Resources Report No. 56. FAO, Rome, pp. 31–54.

EARO (Ethiopian Agricultural Research Organization) (1998) Resource management for improving and sustaining crop and livestock production on highland Vertisols in Ethiopia. Progress Report No. 7. EARO, Addis Ababa. (Mimeo.)

EARO (Ethiopian Agricultural Research Organization) (1999) Resource management for improving and sustaining crop and livestock production on highland Vertisols in Ethiopia. Progress Report No. 9. EARO, Addis Ababa. (Mimeo.)

Goe, M.R. (1987) Animal traction on smallholder farms in the Ethiopian highlands. PhD dissertation, Department of Animal Science, Cornell University, Ithaca, New York, USA.

ILRI (International Livestock Research Institute) (1996) Resource management for improving and sustaining crop and livestock production on highland Vertisols in Ethiopia. Progress Report, ILRI, Addis Ababa. (Mimeo.)

IUCN (International Union for the Conservation of Nature) (1990) *Ethiopian National Conservation Strategy*. Phase I Report. IUCN, Addis Ababa, Ethiopia.

Jutzi, S.C., Anderson, F.M. and Astatke, A. (1986) Low cost modifications of the traditional Ethiopian type plough for land shaping and surface drainage of heavy clay soils: preliminary results from on-farm verifications. In: Starkey, P. and Ndiane, F. (ed.) *Animal Power in Farming Systems*. German Technical Agency, Eschborn, Germany, pp. 127–132.

Kassi, M., Jabbar, M.A., Kassa, B. and Mohamed Saleem, M.A. (1999) Benefits of integration of cereals and forage legumes with and without crossbred cows in mixed farms: an *ex ante* analysis for Highland Ethiopia. *Journal of Sustainable Agriculture* 14(1), 31–48.

Mohamed Saleem, M.A. and Astatke, A. (1996) Options to intensify cropland use for alleviating smallholder energy and protein deficiencies in the East African Highlands. *Field Crops Research* 48, 177–184.

Morgan, W.T.W. (1972) The exploitation of the East-African environment. In: Morgan, W.T.W. (ed.) *East Africa: Its People and Resources*. Oxford University Press, Oxford, UK, pp. 295–302.

Tedla, A. and Mohamed Saleem, M.A. (1992) Cropping systems for Vertisols of the Ethiopian highlands. In: *Reports and Papers on the Management of Vertisols (IBSRAM/AFRICA LAND)*. Network Document No. 1, IBSRAM, Nairobi, Kenya, pp. 55–66.

Tesema, T. and Yirgou, D. (1973) Soil burning (*gye*). Its problem and possible solutions. Progress Report, Debre Zeit. Agricultural Experiment Station, College of Agriculture, Haile Selassie I University, Alemaya, Ethiopia. (Mimeo.)

Tinker, P.B. (1989) *Draught Animal Power Implement for Use on Vertisols*. OD Report, AFRC Engineering, Silsoe, Bedfordshire, UK.

Virmani, S.M. (1988) Agroclimatology of the Vertisols and vertic soils areas of Africa. In: *Management of Vertisols in Sub-Saharan Africa*. Proceedings of conference held at ILCA (Addis Ababa, 1987). ILCA, Addis Ababa, Ethiopia, pp. 44–63.

Westpal, E. (1975) *Agricultural Systems in Ethiopia*. College of Agriculture, Haile Selassie I University and Agricultural University of Wageningen. Agricultural Research Report No. 82 611. Wageningen, The Netherlands.

Indian Vertisols: ICRISAT's Research Impact – Past, Present and Future

14

R.J.K. MYERS AND P. PATHAK

Natural Resource Management Program, International Crops Research Institute for the Semi-Arid Tropics, Andhra Pradesh, India

Key words: India, Vertisols, ICRISAT, agricultural research, watershed management, tillage, socio-economics

Introduction

The Vertisols of India, as with those of the rest of the world, can be likened to a sleeping giant in that the very large potential for high productivity still remains dormant. For generations, they were used, regardless of the total rainfall received, for growing one rabi (post-rainy season) crop per year and they remained fallow for the remainder of the year (El-Swaify *et al.*, 1985). Now, with pressures increasing for intensification of agriculture, they will be required to give higher yields, by converting either to kharif (rainy season) cropping or to growing two crops per year. The International Crops Research Institute for the Semi-Arid Tropics (ICRISAT), soon after its establishment, initiated research aimed at increasing the production from Vertisols, and this effort has continued ever since. This research quickly focused on the need for an approach that recognized that development at the watershed scale offered both an opportunity and a challenge to scientists to produce technologies that took advantage of the potential of Vertisols to give sustainable high yields of rainfed crops, if managed correctly. This chapter outlines the history of this research and its achievements, and then discusses the likely future impacts of the research.

Distribution and Properties

Vertisols and associated soils occur extensively in peninsular India between 8°45′ and 26°0′ north latitude and 66°0′ and 83°45′ east longitude, covering about 72.9 million ha. This is roughly 22.2% of the total geographical area of the country (Murthy *et al.*, 1982). About 80% of the Vertisols in India lie in the states of Maharashtra, Madhya Pradesh, Gujarat and Andhra Pradesh, 13% in Karnataka and Tamil Nadu, and the remainder in other adjoining states. In the central region of the country, the Deccan Plateau, the soils are derived from weathered basalts of the Miocene age, transported materials derived from the basalts mixed to some extent with detritus from other rocks. In other areas, particularly in the south, the soils are also derived from basic metamorphic rocks and calcareous clays. In the west, in Gujarat, they are derived from marine alluvium.

ICRISAT has divided its mandate area into 29 production systems. A production system is defined by the key elements of its farming systems, including commodity production trends and key socio-economic variables. Ten of these production systems occur in India. Vertisols occur in all of these Indian production systems, and they are especially prevalent in the production systems of the semi-arid tropical (SAT) parts of India (Laryea *et al.*, 1998; Table 14.1).

The climate in which these soils occur, extending from semi-arid to sub-humid tropics, is characterized by dry and hot pre-monsoon months (April to June) and a dry mild winter. The mean annual rainfall ranges from 500 to 1500 mm, of which 80–90% is received during the monsoon season from June to September (NBSS and LUP–ICRISAT, 1991). This amounts to 42–77% of the mean annual potential evapotranspiration.

Vertisols are important soils of moderate to high agricultural potential in major regions of India. Traditionally, they were used mainly for sorghum, cotton and millet, but now many crops are grown, including soybean, sunflower, safflower, sorghum, pulses, citrus, rice and bananas.

The dominance of the smectite (montmorillonite) group of clay minerals, which causes the soils to shrink and swell with changes in soil moisture, is responsible for the characteristic features of Vertisols and soils of vertic sub-groups, and for their limitations in use. The soils are difficult to work. They are very sticky when wet and cannot be tilled or cultivated for several days after heavy rain. In the dry season they become very hard and large cracks open at the surface and extend deep into the soil profile. In many areas, because of high sodium and magnesium contents, subsoil porosities and aeration are generally poor and roots of annual crops do not develop.

Table 14.1. Relative importance of Vertisols in the production systems of India (taken from Laryea *et al.*, 1998).

	Production system		
Number	Name	Descriptor	Per cent of area
1	Eastern margins of the Thar Desert	Transition zone from arid rangelands to rainfed, short-season millet/pulses/ livestock	4.0
2	Central/eastern Indo-Gangetic Plain	Sub-tropical lowland rainy and post-rainy season, rainfed, mixed cropping	–
3	Western Indo-Gangetic Plain	Sub-tropical lowland rainy and post-rainy season, rainfed	–
4	Central India	Tropical high-rainfall rainy plus post-rainy season, rainfed, soybean/wheat/chickpea	39.6
5	Eastern India	Tropical, lowland, rainfed/irrigated, rice-based	12.6
6	Saurastra Peninsula	Tropical, lowland, short rainy season, rainfed, groundnut/millet	16.4
7	Eastern Deccan Plateau	Tropical, intermediate rainfall, rainy season, sorghum/cotton/pigeonpea	47.2
8	Western Deccan Plateau	Tropical, low rainfall, primarily rainfed, post-rainy season, sorghum/oilseed	39.3
9	Peninsular India	Tropical, intermediate rainy season, sorghum/oilseed/pigeonpea, interspersed with locally irrigated rice	19.9
10	Eastern India	Tropical, upland, rainfed, rice-based	2.1

The 1970s and the Development of the Vertisol Technology Package

The choice of the ICRISAT-Patancheru site, with its endowment of Vertisols and Alfisols, indicated an expected emphasis on Vertisols, and this did happen. A series of micro-watersheds were set up, and these provided the research focus for a significant proportion of the resource management research during the next 25 years.

The research approach

Since its inception, ICRISAT has been working on the development of suitable research methodologies to arrive at improved farming systems for substantial gains in total food production, in consonance with natural resource conservation. Establishing a technology for cropping Vertisols during the rainy season (elimination of cultivated fallow) has been the major thrust of research aimed at increasing the agricultural productivity of Vertisols under rainfed conditions. Recommended approaches for developing improved technological packages have been formulated for Vertisols in the medium-high, more dependable rainfall zones (750–1250 mm $year^{-1}$).

In 1972, B.A Krantz and Jacob Kampen, along with other scientists, initiated the first research work on Vertisol Management at ICRISAT, Patancheru Center. These scientists had long experience with crops and land management practices. They also collected information from scientists in the Indian national agricultural research system (NARS). Based on this information, a package of practices was identified to overcome the production constraints in SAT Vertisols and at the same time address the soil conservation aspects of kharif cropping. Two major experiments were conducted at the ICRISAT Patancheru Center. First, in 1974, operational-scale research was established on eight experimental Vertisol micro-watersheds. Each of these micro-watersheds was treated with one land management system and had a range of different cropping systems, including the traditional practices of post-rainy season crops of sorghum, chickpea or safflower following a rainy season fallow, and improved practices of rainy season maize followed by post-rainy season chickpea, or sorghum intercropped with pigeonpea. Between 1974 and 1988, 13 land management systems were evaluated along with a range of other management practices. Complete data sets do not seem to exist, but it is likely that data collection included grain and stover yields, economic analysis, soil moisture, runoff and soil loss. Secondly, a 'steps in technology' experiment was initiated in 1976. This assessed systems in terms of genotype (local versus improved), fertilizer (farmyard manure versus chemical fertilizer), management (traditional versus improved) and irrigation (none versus supplementary), but data collected seem to be restricted to yields of grain and stover.

In 1977, attempts were initiated to extend the Vertisol technologies beyond the favourable soil and agrometeorological range of the ICRISAT campus at Patancheru. Two large collaborative projects (Farming Systems I, (FS1), Farming System II (FS2)) were initiated off-campus with Indian Council for Agricultural Research (ICAR) institutions and universities. Under these projects, experiments were conducted on 14 stations of ICAR institutions and universities having annual rainfall ranging from 450 to 1500 mm. As a result of the efforts by Krantz, Kampen and others, a package of practices was developed, which became known as 'Vertisol technology' (Binswanger *et al.*, 1980). This technology was suitable for Vertisol areas having annual rainfall between 750 and 1300 mm and where only one crop is taken every year. Using this package, results obtained from various on-station experiments were highly encouraging (Kanwar *et al.*, 1982; El-Swaify *et al.*, 1985). This technology is based on the concept of a small watershed as the basic resource management and conservation unit. The results indicated that the technology offered the possibility of increasing annual yield of cereal and legume rotations and intercrops to 3–5 t ha^{-1} and of giving profits up to 250% (Table 14.2) on the investment. (Ryan *et al.*, 1982). The improved technology reduced runoff and soil loss (Table 14.3), and reduced the waterlogging or drainage problem.

Table 14.2. Economic performance of the Vertisol technology at ICRISAT: averages of annual performances, 1976–1983. (Source: von Oppen *et al.*, 1985.)

Technology/cropping system	Mean yield ($t\ ha^{-1}$)	Gross returns ($Rs\ ha^{-1}$)	Operational cost ($Rs\ ha^{-1}$)	Gross profits ($Rs\ ha^{-1}$)	CV of gross profits (%)	Marginal rate of return (%)
Improved technology						
Maize–pigeonpea						
Intercrop		6765	2080	4705	28	272
Maize	2.71					
Pigeonpea	1.12					
Maize–chickpea						
Sequence		7021	2757	4264	43	159
Maize	3.20					
Chickpea	1.16					
Sorghum–pigeonpea						
Intercrop		8875	2471	6404	26	304
Sorghum	2.88					
Pigeonpea	1.09					
Traditional technology						
Rainy season fallow						
Post-rainy season						
Sorghum and chickpea		1643	682	961	43	
Sorghum	0.57					
Chickpea	0.72					

It was also expected to generate more rural employment. After this, it was decided to initiate the on-farm trials at various locations in close collaboration with NARS.

The key technologies for Vertisol watershed development

The approach includes the following components:

- Small watershed resource management to conserve rainfall, reduce soil erosion and drain off excess water.
- Provision of field and community drains for effective surface drainage and disposal of excess runoff.
- Cultivation of land immediately after harvest of the post-rainy season crop when soil is not too dry for working.
- Broadbed and furrows (BBF) laid at a slope of 0.4–0.8%. The direction of beds is adjusted so as to minimize topsoil movement; thus only minor land shaping/smoothing is required.

Table 14.3. Annual rainfall, runoff, soil loss and peak runoff rate for a cropped Vertisol with broadbed and furrow system and a traditional monsoon fallow system (1974–1994).

		Broadbed and furrow at 0.6% slope, cropped			Traditional flat, monsoon fallow		
Year	Rainfall (mm)	Runoff (mm)	Peak runoff rate ($m^3 s^{-1} ha^{-1}$)	Soil loss ($t ha^{-1}$)	Runoff (mm)	Peak runoff rate ($m^3 s^{-1} ha^{-1}$)	Soil loss ($t ha^{-1}$)
1974	811	116	0.09	1.30	223	0.22	6.60
1975	1041	162	0.06	1.39	253	0.15	5.21
1976	687	73	0.09	0.98	238	0.16	9.20
1977	585	1	0.01	0.07	53	0.06	1.68
1978	1125	273	0.11	2.93	410	0.15	9.69
1979	690	73	0.08	0.70	202	0.15	9.47
1980	730	116	0.06	0.97	166	0.11	4.58
1981	1126	332	0.11	5.04	435	0.16	11.01
1982	615	10	[a]	0.20	20	[a]	0.70
1983	1000	154	[a]	[a]	288	[a]	[a]
1984	564	11	[a]	[a]	75	[a]	[a]
1985	538	4	[a]	[a]	18	[a]	[a]
1986	586	37	[a]	[a]	114	[a]	[a]
1987	837	118	[a]	[a]	206	[a]	[a]
1988	907	187	[a]	[a]	330	[a]	[a]
1989	1023	237	[a]	[a]	282	[a]	[a]
1990	572	123	[a]	[a]	[a]	[a]	[a]
1991[b]	716	14	0.05	[a]	[a]	[a]	[a]
1992[b]	597	47	0.10	[a]	79	0.15	[a]
1993[b]	758	33	0.09	[a]	116	0.19	[a]
1994[b]	798	53	0.03	[a]	163	0.04	[a]

[a]Data not available.
[b]Runoff values from small plots.

- Use of the bullock-drawn wheeled tool carrier (WTC) and using the bed zone for cropping and the furrow zone for movement of bullocks and wheels of the WTC.
- Use of efficient cropping systems based on agroclimatic conditions.
- Dry seeding of rainy season crops just ahead of monsoon rains.
- Planting of post-rainy season crops in the stubbles of rainy season crops, after shallow cultivation.
- Use of improved seed and adequate amount of fertilizer.
- Timely interculture and weed control.
- Timely plant protection.

- Runoff collection and supplemental irrigation. Good potential generally exists for the collection of excess runoff water and its later use for supplemental irrigation. Significant returns can be gained from the application of small quantities of supplemental water on the post-rainy season crops of wheat, chickpea and the high-value vegetable crops.

The technology specifically emphasized two aspects: (i) water control at levels of watershed, field and crop row (*in situ*) through land treatment techniques; and (ii) an integrated approach to soil, water, nutrient and crop management. Land preparation during summer and dry-sowing ahead of monsoon rains helped to minimize machine and animal traffic when the soil was wet. Further, the wheel and bullock traffic was concentrated mostly in the furrow zone, and the soil in the bed zone remained less compacted and had a lower penetration resistance (El-Swaify *et al.*, 1985).

The science behind these technologies

Table 14.4 shows an operational-scale comparison between yields obtained under traditional and improved systems on research watersheds. The productivity of the improved maize–chickpea and maize–pigeonpea cropping systems was markedly higher than that of the traditional post-rainy season crops of chickpea and sorghum. These increases in grain yields were apparent in all years, even though the rainfall during the cropping period varied from 477 to 1089 mm. The performance of maize during the rainy season was particularly impressive. Von Oppen *et al.* (1985) reviewed the economic performance of the Vertisol technology at ICRISAT during 1976–1984 (Table 14.2). Substantially higher gross returns were achieved by using the improved cropping systems and management practices – Rs 6800–8900 ha^{-1} compared with Rs 1600 ha^{-1} from the traditional system. Though the operational costs were three times as high, gross profits rose from Rs 961 ha^{-1} to Rs 4300–6400 ha^{-1} when the improved technology was employed. This increase in profit was not associated with increased risk as the coefficient of variation in the gross profits was similar for both traditional and improved technologies. In addition to being of direct benefit to crop yields, improved management has been much more effective than traditional management in reducing resource losses by runoff and soil erosion (Table 14.4). Despite all of this, adoption of these technologies was disappointing, and, in retrospect, it appears that expectation of adoption of such packages was unrealistic. Only limited on-farm trials were conducted, though when they were conducted in Maharashtra and Andhra Pradesh states in the late 1970s they tended to confirm the findings of the on-station experiments. The profits from the on-farm trials were substantially higher than profits from traditional farming, yet this did not lead to more widespread adoption, perhaps because there were some crop failures due to early cessation of the rains.

Table 14.4. Grain yields under improved and traditional technologies on a Vertisol at ICRISAT in 21 successive years.

		Grain yield (t ha^{-1})			
		Improved system: double cropping		Traditional system: single crop	
Year	Rainfall during cropping period (mm)	Sorghum/ maize	Sequential chickpea intercropped with pigeonpea	Sorghum	Chickpea
1976/77	708	3.20	0.72	0.44	0.54
1977/78	616	3.08	1.22	0.38	0.87
1978/79	1089	2.15	1.26	0.56	0.53
1979/80	715	2.30	1.20	0.50	0.45
1980/81	751	3.59	0.92	0.60	0.56
1981/82	1073	3.19	1.05	0.64	1.05
1982/83	667	3.27	1.10	0.63	1.24
1983/84	1045	3.05	1.77	0.84	0.48
1984/85	546	3.36	1.01	0.69	1.23
1985/86	477	2.70	0.73	[a]	0.84
1986/87	585	4.45	0.38	0.37	1.27
1987/88	841	4.26	1.35	0.80	0.92
1988/89	907	4.64	1.23	0.61	1.18
1989/90	941	3.76	0.75	1.02	1.10
1990/91	673	3.15	0.83	0.65	0.93
1991/92	719	3.78	1.59	1.44	1.35
1992/93	674	4.04	1.09	1.76	2.11
1993/94	787	4.05	1.17	1.20	1.41
1994/95	807	3.88	1.65	0.82	1.46
1995/96	1121	4.44	1.53	1.20	1.30
1996/97	1028	4.52	2.03	1.41	1.46
1997/98	743	4.81	1.73	1.62	1.16
Mean	799	3.56	1.17	0.83	1.06
SD	191	0.71	0.39	0.39	0.41
CV (%)	24	20	34	47	39

[a]No crop sown.

The 1980s – Village-level Testing, Controversy and Multidisciplinary Approaches

This period was significant for attempts to test the Vertisol and watershed technologies in real farming situations, and it was also a time when questions were raised regarding the interpretations of experiments that led to the identification of the watershed technology (Anders and Sharma, 1993).

Village-level testing

To test the performance of the improved technology, on-farm trials were conducted from 1981 to 1984 at various locations. These trials were conducted in collaboration with the State Department of Agriculture, ICAR institutions and universities. Initially, most of the on-farm trials were managed mainly by ICRISAT staff. Only later, the agriculture departments of Andhra Pradesh, Karnataka, Madhya Pradesh and Maharashtra began further testing of the technology on their own initiative. But, in all these trials, the involvement of farmers was very slight.

Results obtained from the various on-farm trials were mixed (Sarin and Ryan, 1983). Highly encouraging results were obtained with the improved technology in Taddanpally, Sultanpur and Begumgunj, which have Vertisols with medium to high annual rainfall (750–1200 mm). At Kanzara and Shirapar, the performance of the improved technology was unimpressive compared with traditional farmers' practices. These two locations did not have the same assured rainfall as the other locations.

Controversy

M.M. Anders joined ICRISAT at a time when there was concern about the disappointing degree of adoption of the Vertisol technology package. He took a keen interest in the topic and explored the thesis that the technology was flawed, and that, in part at least, this was due to misinterpretation of the results of the early experiments. Anders' critique of the early Vertisol watershed experiments (Anders and Sharma, 1993) generated a lot of heat, but not as much productive discussion as might have been hoped. There seems to have been no documentation of any such productive debate if it did occur, and now, with the departure of almost all the key scientists, this unfortunately is not likely to occur.

Multidisciplinary approaches

In 1989, M.M. Anders attempted to deal with these issues by modifying the design of the treatments in three of the ICRISAT-Patancheru microwatersheds. In the revised design, he attempted to deal with statistical design problems, to gather new information on crop nutrition issues and options for crop rotations and to obtain feedback on the limitations to adoption. The new range of treatments included irrigation treatments to determine the degree to which water was limiting performance, farmyard manure (FYM) with a control treatment to evaluate the value of FYM, and there was also a zero fertilizer set of treatments to quantify the effects of the fertilizer treatments. These new treatment combinations were to continue until 1997.

In 1982, T.J. Rego established an experiment in one of the Vertisol micro-watersheds aimed at evaluating the sustainability of crop rotations, and in particular to evaluate the effects of rotations and inputs on soil fertility. This experiment was to run until 1999.

The 1990s – New Directions and Potential Impact

Socio-economics – analyses of success and failure, and more impact studies

From 1994 to 1997 visits were made to all locations where Vertisol technology on-farm trials were conducted. The main objectives of these visits were: (i) to undertake the appraisal on adoption of different components of the Vertisol technology at various locations; and (ii) to get feedback on the limitations for non-adoption. Some of the outcomes are given below.

Broadbed and furrow (BBF)

- Farmers were by now convinced of the benefits of the BBF technology. However, they still faced constraints in adopting the BBF method for draining the excess water from their fields.
- High cost of equipment used for making BBF was the main reason for non-adoption of BBF.
- Lack of suitable tractor-drawn equipment for making BBF inhibited use of BBF in Madhya Pradesh, where tractors had replaced bullocks.
- In areas where waterlogging is a serious problem, farmers are using the basic concept of BBF and make furrows between rows during interculture operation. They found this modified system more convenient to adopt than that previously recommended in the Vertisol technology package.

Field and main drains

Farmers have generally recognized the value of field drains and main drains for disposal of excess runoff. It was observed that many farmers had constructed field drains required for good drainage within their fields. However, the concept of main drains or community drains has not caught on and individual farmers are not willing to participate when drains are located outside their fields or involve the drainage flows of many farmers. However, the main drains constructed by ICRISAT still exist, although they are not well-maintained.

Crops and cropping pattern

At one of the locations, soybean is now the most important rainy season crop, occupying about 73% of the total cultivated area. Farmers acknowledged the contribution of ICRISAT in introducing soybean to the village, even though soybean was not an ICRISAT mandate crop, and was not part of the Vertisol technology package.

- About 30% of the cultivated area was under soybean plus pigeonpea intercropping, which covers both rainy and post-rainy season. Before introducing the Vertisol technology in the village, lands were largely cultivated during the post-rainy season after rainy season fallow. Farmers acknowledge the role of ICRISAT in introducing the intercropping.
- At another location, crops and cropping systems suggested by ICRISAT were not taken up. Farmers had reverted back to their old crops and cropping systems.

Dry seeding

Dry seeding did not gain much popularity. Few farmers in Maharashtra use dry seeding in cotton-based systems. Only about 2% of farmers were following this practice with soybean. Whereas the widespread adoption of soybean and cotton throughout the Vertisol region was not anticipated, the subsequent non-adoption of dry seeding was to be expected. Observations at ICRISAT had shown that high soil temperatures with dry seeding resulted in reduced germination and establishment of crops with oil-rich seeds. Thus the farmers were correct in rejecting dry seeding on crops for which it was unsuited. Inability to use dry seeding provided a new obstacle to the implementation of double-cropping, since dry seeding of the kharif crop was critical and, in its absence, new farmer-acceptable shorter duration cultivars would be necessary.

Several constraints in the adoption of this practice were reported.

- Where there is excess rainfall, the probability of seed damage is very high.
- Early-sown fields face the problem of free cattle grazing.
- No reliable predictions are yet available on the arrival date of rainfall.

Fertilizer application

Most of the farmers apply fertilizers to soybean, pigeonpea, wheat and chickpea. Most of the farmers also apply FYM, although use of FYM is declining because of its scarcity in the area. The rates used by the farmers are different from those suggested by ICRISAT. Also this practice of applying fertilizer existed even before the on-farm trials were conducted.

Further evaluation from long-term experiments

We have completed an analysis of physical and chemical changes that have occurred in treatments of some of the long-term experiments on Vertisols. K.M.A. Kendaragama (personal communication) conducted a pilot study to examine some of the analytical methods that would be useful in identifying soil chemical changes in these Vertisols. He sampled an experiment on Vertisols that had been running for 8 years, and examined a small number of the more extreme cropping systems (maize followed by chickpea rotated with maize followed by safflower; soybean/pigeonpea (intercropped) rotated with millet followed by safflower and inputs (traditional FYM inputs); improved fertilizer applications) and tested some optional analytical procedures (Walkley–Black organic C; Nelson–Summers organic C; total N; ammonium N; nitrate N; mineralizable N; bicarbonate P). He found from this study that the total C analysis identified treatment differences that were not identifiable with the Walkley–Black incomplete combustion method (soybean/pigeonpea rotated with millet followed by safflower was accumulating organic matter), also that mineralizable N was higher in soil from the FYM treatments whereas the fertilizer-input treatment showed no differences, and that bicarbonate P was a poor indicator of P fertility in this Vertisol (K.M.A. Kendaragama, personal communication) (Table 14.5).

Interest in vertic Inceptisols

Vertic Inceptisols cover about 60% of the area of production systems 7 and 8 in India. They often have slopes exceeding 2%, are shallow in depth and suffer from erosion. The Vertisol technology package, developed as it was at ICRISAT's Patancheru location, was judged to have addressed mainly the deeper Vertisols and the rainfall areas of 800 mm or more. The package was not necessarily appropriate to the very substantial areas of vertic Inceptisols in the region. Thus, in 1996, the BW7 micro-watershed was developed to provide new information on farmer-acceptable management options for vertic soils. This experimental area is now established to evaluate soybean-based systems as a reflection of the large change that has occurred in the Indian Vertisol areas since the establishment of the earlier experiments. Included in the experimentation is examination of the water balance and nutrient balance for soybean-based systems on vertic Inceptisols, and research conducted jointly with Michigan State University to develop a digital terrain modelling capability.

Watersheds in the real world – socio-economic considerations

Adolph (1997) summarized the status of the watershed 'industry' in India. This 'industry' consists of a large number of projects conducted by governmental agencies and non-government organizations (NGOs). Whereas some successes have occurred, the overall impact was not encouraging. Farmers had refused to adopt recommended practices for a number of reasons, or had abandoned them once project support was withdrawn. Where there had been success, three key factors were identified, all being characteristics of participatory approaches: (i) farmers' involvement in the choice of soil and water conservation technologies, as well as the incorporation of indigenous practices into project design; (ii) farmers' contribution (in cash or labour) to

Table 14.5. Soil organic C, N and P status indicators of a Vertisol as affected by crop rotation and inputs in a long-term experiment.

Rotation	Fertilizer	Depth (cm)	Organic C (%)			N (mg kg^{-1})			P (Bicarbonate) (mg kg^{-1})
			Walkley–Black	Nelson–Summers	Total	NH_4	NO_3	Mineralizable N	
MCMS[a]	Traditional	0–15	0.43	0.53	534	1.6	1.9	8.4	2.0
	FYM	15–30	0.38	0.53	455	1.8	1.1	5.0	1.4
	Mineral	0–15	0.41	0.58	522	2.0	1.8	9.1	4.9
	fertilizer	15–30	0.35	0.52	400	1.4	0.8	5.9	1.6
SPMS[b]	Traditional	0–15	0.51	0.68	643	1.7	1.1	18.6	2.3
	FYM	15–30	0.41	0.49	450	3.1	0.4	7.7	1.0
	Mineral	0–15	0.49	0.64	643	2.3	2.0	13.5	2.4
	fertilizer	15–30	0.36	0.48	448	2.3	0.6	7.8	1.1
CV%			10	11	11	32	26	21	32
SEcfd[c]			0.023	0.030	30.0	0.321	0.217	1.374	0.373
Crops			0.023	0.029	23.5	0.306	0.267	0.967	0.292
Fertilizers			0.024	0.035	31.5	0.376	0.181	1.149	0.379
Depth			0.023	0.029	23.5	0.306	0.267	0.967	0.292

[a]MCMS, maize–chickpea–maize–safflower.
[b]SPMS, soybean/pigeonpea–millet–safflower.
[c]SEcfd, standard error (cropping system × fertilizer × soil layer).

the programme costs; and (iii) existence and functioning of local organizations. It is worrying that no ICRISAT natural resource management scientists were involved in this study. Nevertheless, ICRISAT is incorporating these lessons into its new watershed research and development studies.

Watersheds in the real world – the Ethiopian project

In partnership with the Ethiopian NARS and International Livestock Research Institute, ICRISAT is conducting a modest but challenging project on Vertisol management. This project appears to be a successful application of the watershed approach in Africa.

Watersheds in the real world – pilot study at Zaheerabad

When ICRISAT finally commenced research in a real-world watershed in 1997, it was not a Vertisol system. However, this study near Zaheerabad, Andhra Pradesh, which involved collaboration with an NGO, may have helped to initiate the work that soon followed. In the Zaheerabad watershed study, ICRISAT set about monitoring the activities and interventions initiated by the NGO, and the impact of those activities on the water balance and productivity of the system. The location has also provided an opportunity to develop new tools to assist watershed development. Thus a new method that combines geographical information systems (GIS) with remote sensing has been developed to enable more rapid survey of watersheds, and new watershed development manuals have been prepared for distribution to watershed development teams. This work is now in its third year and will be subject to review in late 1999. An attempt to obtain funding from an NGO/donor for similar monitoring and improvements of intervention strategies in Vertisol watersheds was unsuccessful – the donor believed that the cost was too high, and the conclusion is that they are prepared to continue what they are doing without knowing the impact of their work.

Watersheds in the real world – the future: watershed development in India, Thailand and Vietnam

In 1999, ICRISAT commenced a new project on watershed development, in which it works with NARS, NGO and International Agricultural Research Center partners in India, northeast Thailand and northern Vietnam. In India, the target watersheds are all Vertisols or vertic soils. Success in this project will chart the future of ICRISAT's further involvement in natural resource management research in Asia, and will also provide a potential springboard for development of the watershed approach in Africa.

New tools

As mentioned above, we are in the process of bringing out two watershed development manuals that cover the engineering aspects of watershed development (P. Pathak and M.C. Klaij). Also, Pathak has developed a new methodology that combines GIS and remote sensing to assist in the surveying and laying out of watersheds. In addition, new work has been initiated on improving the efficiency and benefits of groundwater recharge interventions.

Soil and crop modelling may now be about to deliver some of its long-promised benefits. The development of digital terrain modelling in collaboration with Michigan State University should provide a very useful tool for watershed development. Similarly the Agricultural Production Systems Simulator (APSIM) modelling package offers a new and powerful tool for assisting farmers in the SAT in their decision making. Now that pigeonpea, millet, phosphorus and manure modules have been incorporated into APSIM, it has increased potential for use in the SAT.

Conclusions and Lessons Learnt

- Sustainability is possible only through people's participation.
- To ensure people's participation, cost sharing is a must. It has been observed that there is no effective participation without cost sharing.
- To improve the people's participation, project priorities have to be demand driven (not supply driven) with sufficient flexibility.
- The project should be programme- and process-oriented and not target-oriented. The participatory process takes a back seat when there is emphasis on the physical and financial targets.
- The capacity building for the project staff for technical competence and for all the stakeholders from NGOs to beneficiaries for institutional development should start from the planning stage and should be a continuous process.
- Evolution of an organization culture: since watersheds are large community-connected projects with the potential to create serious conflict situations, creation of an appropriate organization culture is very important.
- The role of women is crucial in watershed management. Women have made a significant contribution in several watershed projects.
- The landless are important components of watersheds, and can contribute to the problem of poverty and degradation, as well as potentially contributing to the solutions.
- Monitoring and evaluation, starting from the second year, followed by every year, should be part of the project activity.
- If poverty alleviation and equity are the overall objectives of the project, the scope of the project should be widened to include infrastructure

facilities and other support activities based on local potential, and the approach should be the development of a defined area on a watershed basis. In the long term this can be a significant contribution to the size of the watershed project.

After 26 years, ICRISAT has made substantial achievements in the area of germplasm enhancement of mandate crops. Its achievements in natural resource management research are rather less evident, and this applies to Vertisols as much as to other soils. The impact of the high-profile Vertisol/ watershed technologies has been slow to arrive and several comments are relevant.

First, the Vertisol technology package was designed at a time when it was still assumed that developed technologies would be handed over to NARS and adoption would naturally follow from traditional extension. With the benefit of hindsight, the general failure of traditional extension in dryland agriculture is now well recognized.

Second, the Vertisol technology package was designed and promoted as a 'one size fits all' package, whereas it suited best the true Vertisols and the high end of the range of the SAT where drainage improvement by use of BBF could be expected to be beneficial.

Third, the huge, unanticipated expansion of soybeans (a non-mandate crop for ICRISAT) into the Vertisols and vertic Inceptisols of the Indian SAT largely invalidated some key components of the Vertisol technology package. In particular, soybeans were not suited to dry seeding, and this meant that double-cropping would not be an option unless acceptable shorter-duration cultivars became available.

References

Adolph, B. (1997) *Scaling up Participatory Approaches to Watershed Management: Challenges and Opportunities.* Summary proceedings of the regional workshop, 8 and 9 January 1997, Krishi Vigyana Kendra, Zahirabad, Andhra Pradesh, India. Integrated Systems Project Report Series No. 7. International Crops Research Institute for the Semi-arid Tropics, Patancheru, Andhra Pradesh, India, and University of Hohenheim, Stuttgart, Germany, 30 pp. (Semi-formal publication.)

Anders, M.M. and Sharma, S.K. (1993) *An Historical Perspective of Vertisol Watershed Research at ICRISAT 1975–1989: Steps in Technology and Operational-scale Demonstration Studies.* ICRISAT Resource Management Program, Production Agronomy Progress Report II. International Crops Research Institute for the Semi-arid Tropics, Patancheru, Andhra Pradesh, India.

Binswanger, H.P., Virmani, S.M. and Kampen, J. (1980) *Farming Systems Components for Selected Areas in India: Evidence from ICRISAT.* Research Bulletin No. 2. International Crops Research Institute for the Semi-Arid Tropics, Patancheru, Andhra Pradesh, India, 40 pp.

El-Swaify, S.A., Pathak, P., Rego, T.J. and Singh, S. (1985) Soil management for optimized productivity under rainfed conditions in the semi-arid tropics. *Advances in Soil Science* 1, 1–61.

Kanwar, J.S., Kampen, J. and Virmani, S.M. (1982) Management of Vertisols for maximizing crop production – ICRISAT experience. In: *Proceedings of the 12th International Congress of Soil Science, 'Vertisols and Rice Soils of Tropics'*, 8–16 February 1982, New Delhi, India. Indian Society of Soil Science, Indian Agricultural Research Institute, New Delhi, India, pp. 95–118.

Laryea, K.B., Bantilan, F.T., Mohan Rao, P., Moinuddin, M. and Pathak, P. (1998) *Distribution of Soils in Production Systems in India.* Information Bulletin No. 53. International Crops Research Institute for the Semi-Arid Tropics, Patancheru, Andhra Pradesh, India, 28 pp.

Murthy, R.S., Bhattacharjee, J.C., Lansey, R.J. and Pofali, R.M. (1982) Distribution, characteristics and classification of Vertisols. In: *Proceedings of the 12th International Congress of Soil Science, Vertisols and Rice Soils of the Tropics*, 8–16 February 1982, New Delhi, India. Indian Agricultural Research Institute, New Delhi, India, pp. 3–22.

NBSS and LUP–ICRISAT (1991) *The Suitabilities of Vertisols and Associated Soils for Improved Cropping Systems in Central India.* NBSS and LUP, Nagpur and ICRISAT, Patancheru, India, 61 pp.

Ryan, J.G., Virmani, S.M. and Swindale, L.D. (1982) Potential technologies for deep black soils in relatively dependable rainfall regions of India. In: *Proceedings of the Seminar on Innovative Technologies for Integrated Rural Development*, 15–17 April 1982. Indian Bank, New Delhi, India, pp. 41–62.

Sarin, R. and Ryan, J.G. (1983) *Economic Assessment of Improved Watershed-based Technology Options in On-farm Experiments.* Economics Program Report No. 46. International Crops Research Institute for the Semi-Arid Tropics, Patancheru, Andhra Pradesh, India, 20 pp. (Limited distribution.)

von Oppen, M., Ghodake, R.R., Kshirsagar, K.G. and Singh, R.P. (1985) Performance and transfer constraints of improved Vertisol technology. Reprinted from: *Entwicklung und Landlicher Raum* 19(6/85), 11–14.

Planning and Facilitating a 'Negotiated Learning and Action System': Participatory Research to Improve Soil Management Practices on Indian Vertisols and Alfisols

15

C.A. KING,[1] H.P. SINGH,[2] G. SUBBA REDDY[2] AND D.M. FREEBAIRN[3]

[1]*Department of Primary Industries, Toowoomba, Queensland, Australia;* [2]*Central Research Institute for Dryland Agriculture, Santoshnagar, India; and* [3]*Queensland Department of Natural Resources, Toowoomba, Queensland, Australia*

Key words: India, Vertisols, Alfisols, soil management, farmers, rainfall simulators, participatory research, on-farm research, action learning

Introduction

Participatory research approaches are being used to increase the involvement of the beneficiaries in development and research activities. This has been due to awareness that more traditional approaches have often led to inappropriate, irrelevant and unequally distributed technologies and unrepresentative decision making. As research initiatives began to include farmers, institutional researchers and practitioners became more interested in indigenous knowledge and innovation. Linking scientific and indigenous forms of experimentation proved useful in enhancing the quality of research and farmer knowledge and experience provided context and meaning to the research agenda. Approaches such as farming systems research and participatory technology development were developed to reflect a systems approach and involve farmers in research activities. Today, the current trend in agricultural and resource management extension is for practitioners to facilitate and use more participatory learning processes with farmers and scientists.

The need for participatory research in Vertisol and Alfisol management

Vertisols and Alfisols occupy nearly 62 million ha of India's 143 million ha of cultivated land. Rainfed agriculture in India constitutes 70% of this cultivated area, provides 44% of food grain production and supports 40% of the population. In these rainfed areas, 80% of farming families are smallholders (including lease farmers) and nearly 80% of India's 600 million livestock are located. There is a need to reduce soil degradation in these areas where over half of the cultivated land suffers from soil erosion, waterlogging and salinity problems. At present, there is an impending urgency to upgrade the productivity of rainfed lands in the next decade from 0.8 t ha^{-1} to an estimated 2.0 t ha^{-1} to support increasing population needs.

The rainfed agricultural areas of India are characterized by resource-poor and marginal farmers, with poor infrastructure and low investments in technology and input. Conventional research and extension, however, have been based on top-down transfer of technology from research institute to farmer and, as a result, have provided technology that only suited the homogeneous resource conditions of highly productive farming systems with easy access to inputs and services. Researchers have since been questioned by development practitioners and agencies about their approaches and urged to develop technologies that are more relevant to complex environments. Participatory research provides a mechanism for research institutes and extension agencies to achieve this.

The Australian Centre for International Agricultural Research (ACIAR) Project 9435, through participatory research, aims to produce tools and indicators for planning more sustainable soil management practices in these rainfed areas. The project is a collaborative project between three main agencies including the Department of Natural Resources (DNR) and the Department of Primary Industries (DPI) in Queensland, Australia, and the Central Research Institute for Dryland Agriculture (CRIDA), India. It is an initiative of ACIAR and the Indian Council for Agricultural Research (ICAR). In India, research efforts are under way in three villages, including Nallavelli (near Hyderabad), Madabhavi (Bijapur District) and Pampanur (Anantapur District) and three closely located CRIDA centres, including Hyderabad (Andhra Pradesh), Bijapur (Karnataka) and Anantapur (Andhra Pradesh). The project has an overwhelming sense of complexity where interdisciplinary research is taking place with farmers, scientists and extension agents from both Australia and India using adult learning and action research processes to explore innovations and facilitate change.

This chapter illustrates a 'negotiated learning and action system' research process based on the notions of participatory learning and systems thinking. The authors' aim is to provide the reader with insights gained from planning and facilitating this learning system.

Methodology

An action research methodology was used as the basis for both researching and improving the research process. Action research, as the name suggests, has the dual aims of action and research. The aim of action is to bring about change in a community, organization or programme. The aim of research is to increase understanding on the part of the researcher and in some cases the community, organization or programme.

One of the most simplified and easy to apply forms of action research is that developed in 1992 by Zuber-Skerrit (1995), where the core of the action research process is based upon a spiral of cycles of action and research consisting of four phases. These are planning, action, observation and reflection. Planning consists of analysing a complex situation and developing a strategic action plan. Action is implementing the plan – the practical testing phase. Observing is monitoring the action taken. Reflecting is the practice of evaluating the results over the whole process. This final stage of reflecting is perhaps the most critical part in the process as it allows for continual refinement.

There are benefits of action research over more traditional research methods, particularly in the case where the researcher also acts as a change agent. Dick (1993) notes that action research 'offers the opportunity for practitioners to achieve better research outcomes from their practice without undermining the practice'. It does this by placing emphasis on responsiveness and specificity. More conventional research places its emphasis on replication and generalizability. That is, there are trade-offs involved when choosing a research approach. In a change programme both responsiveness and specificity are given precedence over replication and generalizability to provide better opportunities for local change. Dick (1993) suggests that 'if you can achieve [action research] in a way which allows for some replicability, so much the better'.

In order to illustrate the components of action research, this chapter reports selected parts of the participatory research in terms of the planning, actions, observations and reflections that took place during the research process. This challenges more traditional styles of reporting, which are often represented as methodology, results, discussion and conclusions. The authors' aim is to take you through this alternative style as you read the chapter to enable you to anticipate and interpret situations and learnings during the research process. As a guide, an action research cycle is positioned at the beginning of each of the four components to illustrate where you are in the research process.

The Planning (Analysing a Complex Situation and Designing a Strategic Action Plan)

Analysing a complex situation

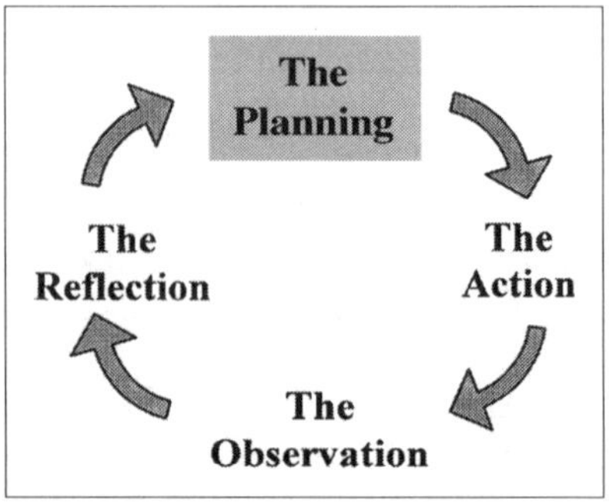

To analyse the situation, qualitative data were collected at each of the sites using a focus group methodology (Krueger, 1988). This methodology is a widely used research tool in social science, where individuals respond in their own words, using their own categorizations and perceived associations (Stewart and Shamdasani, 1990). Focus groups differ from regular discussion groups because a moderator is present to provide all participants with every opportunity to have input and enable participants to lead the direction of the discussion until no more new ideas are raised. The moderator asks a series of open-ended key questions and refrains from introducing ideas.

The key questions focused primarily upon participants' perceptions of sustainability, sustainable farming systems and farm management practices. The outcome from the study was an enhanced understanding by the project team about the similarities and differences between farmers' and scientists' perceptions about sustainability, sustainable farming systems, the types of indicators used to assess more and less sustainable systems and how farmers make decisions with regard to farm management. Details of the findings can be found in Progress Report (1997). Focus groups also had a secondary function, which was to create rapport between team members and farmers and provide a mechanism for farmers to learn from each other during the focus group process. Previous research was also used to inform our understanding of the situation.

Designing a strategic action plan

To design the participatory research process a '5P framework' was developed consisting of *paradigm, purpose, principles, people* and *process*. Before using the framework the first four of these Ps had to be explored and established by the authors. They are described below.

5 Ps – the paradigm

To explore our paradigm, a general description provided by Patton (1990) was used. After much discussion, the words that best described our paradigm were 'facilitators of change'. The assumptions behind this are that change should:

(i) occur with regard to managing resources; and (ii) be facilitated rather than enforced or determined.

> A paradigm is a world view, a general perspective, a way of breaking down the complexity of the real world. As such, paradigms are deeply embedded in the socialisation of adherents and practitioners: paradigms tell them what is important, legitimate, and reasonable. Paradigms are also normative, telling the practitioner what to do without the necessity of long existential or epistemological considerations. But it is this aspect of paradigms that constitutes both their strength and their weakness – their strength in that it makes action possible, their weakness in that the very reason for action is hidden in the unquestioned assumptions of the paradigm.
>
> (Patton, 1990, p. 37)

5 Ps – the purpose

The purpose was defined as the reason behind the research process. The main aims and objectives of the ACIAR project were re-examined. We concluded that the two main purposes of the research process were:

1. To research more sustainable soil management practices suitable for use by farmers.
2. To create a change in the knowledge, aspirations, skills or attitudes of those involved in the research to improve their decision making or practice with regard to more sustainable farming.

5 Ps – the principles

The question we used to establish the principles was 'What are the key beliefs we would like to uphold in the research process?' This was a challenging question and, during the exploration of ideas, three key principles emerged. These were negotiated learning, action research and systems thinking. These are described below.

NEGOTIATED LEARNING (AS A FORM OF INTERACTIVE PARTICIPATION) A number of reasons were identified as to why we thought participation was important in the research process. These were: (i) everyone has a right to be part of any decision making that affects their livelihood; (ii) to develop appropriate and relevant technology, and innovation end users should be involved; and (iii) people should have the ability to choose whether to participate or not. After considering and exploring different types, issues and notions of participation, a key principle established was one of negotiated learning. The main ideas behind this principle were:

- Farmer knowledge and traditional scientific knowledge are both valid and complementary sources of wisdom in the research process.
- People should participate in joint analysis, development of action plans, and formation or strengthening of local groups or institutions.
- Participation is seen as a right, not just the means to achieve project goals.

- Learning methodologies should be used to seek multiple perspectives and groups determine how available resources are used.
- Different realities exist among participants and the process requires ways for these participants to inform each other and come to some negotiation (not consensus) about the learning that takes place and actions to be taken.
- Shared learning is a critical step for all participants in negotiation so that everyone has opportunities to explore issues and solutions and be part of the construction of knowledge.

To establish this principle we used a number of mechanisms to articulate the type of participation we wanted to facilitate during the research process. To aid this, a typology of participation by Pretty (1995) was useful and can be seen in Table 15.1. The typology that we felt best described our principle was interactive participation.

ACTION RESEARCH Action research was selected as a key principle because we wanted to facilitate change within the villages and the organization as well as improve the understanding of the participants about more sustainable soil management practices. As mentioned earlier, the action research process is based upon a spiral of cycles of action and research consisting of four phases. These are planning, action, observation and reflection. Planning consists of analysing a complex situation and developing a strategic action plan. Action is implementing the plan – the practical testing phase. Observing is monitoring the action taken. Reflecting is the practice of evaluating the results over the whole process. This final stage of reflecting is perhaps the most critical part in the process as it allows for continual refinement. The action research process has been shown to provide an effective means of researching real solutions to real problems in real-life situations.

Action research embodies Kolb's experiential learning cycle, which consists of four stages, including concrete experience, reflective observation, abstract conceptualization and active experimentation (Kolb, 1984). The learning cycle begins with the learner having an experience (*concrete experience*). The learner then reflects on this experience (*reflective observation*) and creates generalizations that integrate these reflections into theories (*abstract conceptualization*). These generalizations and hypotheses are then tested in more complex situations, through action (*active experimentation*). This process leads to new concrete experiences and another cycle evolves (Fig. 15.1).

The literature provides an array of different perspectives and perceptions of the processes involved in action research. Action research challenges the more conventional scientific paradigms, but provides a more practical and ethical approach to problem solving and instigating change in real-life situations.

SYSTEMS THINKING We chose a systems approach as it has been shown to be useful because it 'takes on a holistic view of the world and allows for interactions to be discovered' (Roling and Jiggins, 1998). It is particularly useful in dealing

Table 15.1. Typology of participation (from Pretty, 1995).

Typology	Characteristics of each type
Manipulative participation	Participation is simply a pretence, with 'people's' representatives on official boards but who are unelected and have no power
Passive participation	People participate by being told what has been decided or has already happened. It involves unilateral announcements by an administration of project management without any listening to people's responses. The information being shared only belongs to external professionals
Participation by consultation	People participate by being consulted or by answering questions. External agents define problems and information gathering processes, and so control analysis. Such a consultative process does not concede any share in decision making, and professionals are under no obligation to take on board people's views
Participation for material incentive	People participate by contributing resources, for example, labour, in return for food, cash or other material incentives. Farmers may provide the fields and labour, but are involved in neither experimentation nor the process of learning. It is very common to see this called participation, yet people have no stake in prolonging technologies or practices when the incentives end
Functional participation	Participation seen by external agencies as a means to achieve project goals, especially reduced cost. People may participate by forming groups to meet predetermined objectives related to the project. Such involvement may be interactive and involve shared decision making, but tends to arise only after major decisions have already been made by external agents. At worst, local people may still only be co-opted to serve external goals
Interactive participation	People participate in joint analysis, development of action plans and formation or strengthening of local institutions. Participation is seen as a right, not just the means to achieve project goals. The process involves interdisciplinary methodologies that seek multiple perspectives and make use of systemic and structured learning processes. As groups take control over local decisions and determine how available resources are used, so they have a stake in maintaining structures or practices
Self-mobilization	People participate by taking initiatives independently of external institutions to change systems. They develop contacts with external institutions for resources and technical advice they need, but retain control over how resources are used. Self-mobilization can spread if governments and non-governmental organizations provide an enabling framework of support. Such self-initiated mobilization may or may not challenge existing distributions of wealth and power

with complex situations. Through more traditional reductionist approaches, interactions are neglected as research focuses around exploring and analysing separate parts of the system.

Taking on a systems approach at the farm and catchment level has involved exploring the complexity of interactions within the 'hard' and 'soft'

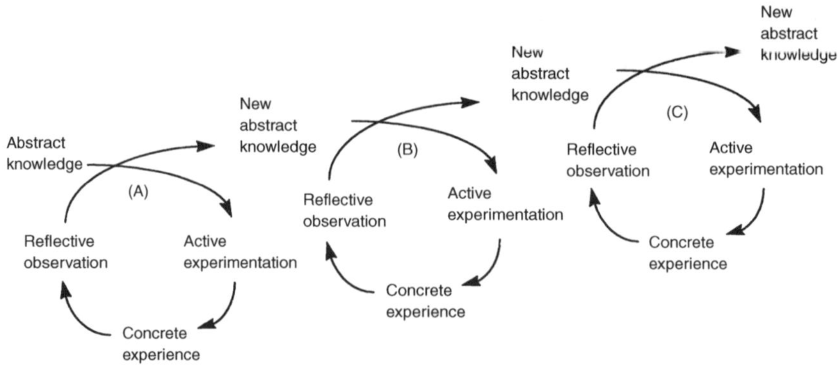

Fig. 15.1. A series of experiential learning cycles (A, B and C) with increasing knowledge as the process continues (adapted from Kolb, 1984).

systems. The hard system can be viewed as the biophysical components, which can be modelled, particularly by simulation at the farm and catchment level. The soft system can be viewed as the interactions between the biophysical components, technology and the farm family at the farm and catchment level. If a 'soft' systems approach is used (i.e. recognition that people have different realities), a system becomes a construct in which to understand and describe the interactions and relationships within a constructed boundary and a system becomes a system of inquiry.

In our project, the hard system would include the exploration of the biophysical components of the field plots and related catchment. The soft system would include the researchers, scientists, extension agents and farmers and their interactions with the hard system components.

5 Ps – the people

There was a need to identify the key people to be included in the research process. In this project, key scientists to be involved in the project were already determined. We did, however, have the opportunity to select villages that we felt would be appropriate. We realized that it would be beneficial to view and facilitate the scientists and farmers involved in the research as an ongoing learning group. The number of ongoing learning groups focusing on resource management issues has increased substantially in the past 5 years. Working with groups can be more efficient than one-on-one activities in terms of allocated resources, has the potential to explore more complex issues than one-off learning activities, and provides opportunities for individuals (in a group environment) to build upon previous knowledge in a purposeful and strategic manner.

King *et al.* (1998) raise the point that ongoing learning groups often have fixed membership and suggest that, as new issues emerge through continuous learning approaches, relevant stakeholders (with respect to these new issues) are often excluded. This is particularly important when it comes to the decision-making process. Ongoing learning groups that have fixed members over time underestimate the dynamic nature of these learning systems, particularly in terms of knowledge use and exchange to identify and address wider system issues. For this reason, we also wanted to enable the group to be flexible enough so that we could introduce relevant people at different times during the process.

5 Ps – constructing the process through iteration

To use the framework, three main steps were taken. First, the *paradigm*, *purpose* and *principles* were placed into the framework. Second, relevant *people* for the research process were selected in light of these three Ps. Third, through a process of iteration between these four Ps, the *process* was constructed. Once the initial framework was complete (Table 15.2) it provided a mechanism for periodic reflection and continuous improvement. That is, it could be modified through the course of the project in response to new insights and learnings. For example, as learning occurs, perceptions of the issues may change and different people may enter and leave the process. The participatory research process is simplified in this table and is described in more detail in the following section.

After completing the planning of the 'new' participatory research process, our role in the process became clearer. Our role would be to *facilitate* a '*negotiated learning and action system*' where the system comprised scientists, farmers and other extension agents in a process of learning and research about more sustainable soil management.

Table 15.2. A simplified version of the '5P framework' at the initial design stage of the research process.

Paradigm	Principles	Purpose	People	Process
Facilitating change	• Negotiated learning • Action research • Systems thinking	To research better surface management practices To create change in terms of the knowledge, aspirations, skills, attitudes and practice (KASAP) of all involved in the research process	Scientists • Biophysical • Economic • Social Farmers • Women • Men	Facilitating a 'negotiated learning and action system' • Soil plot experiments using a rainfall simulator • On-farm and/or on-station soil plot field trials

Actions, Observations and Reflections

The 'negotiated learning and action system' facilitated during the ACIAR Project encompassed two types of soil plot experiments. The first of these involved the use of field plot trials as a tool for researching different soil management options over the cropping season. The second was the use of a portable rainfall simulator as a tool for researching the effect of rainfall intensity on different soil management options. To provide you with an example of what actually happened when we facilitated these, we have included two illustrations (Fig. 15.2 and Fig. 15.3) of what occurred in Nallavelli village (Hyderabad) during the course of these two research activities. From these figures, we have selected components to illustrate some of the actions, observations and reflections made during the research process. These are shaded. That is, in the first example (using on-farm and on-station research plots) only part of the process is described and in the second example (using a rainfall simulator) the entire process is presented.

Empirical Line 1: Exploration with the Farm Family, Village Farmers and Scientists about their Previous Research Approaches, Negotiating a New Learning Approach and Seeking Involvement (see Fig. 15.2)

The Action (facilitating the 'negotiated learning and action system')

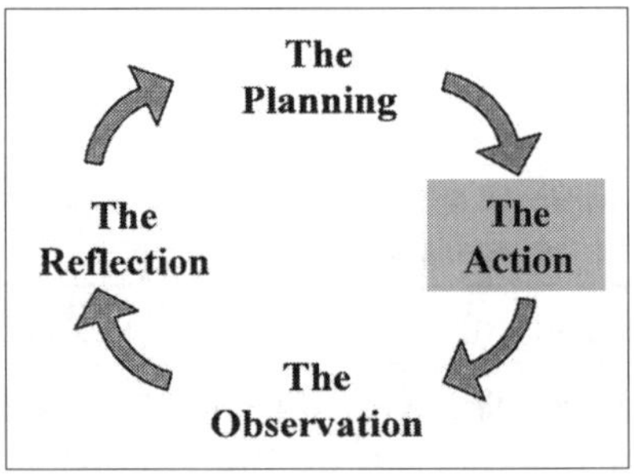

We tried to ensure that farmers and scientists who could potentially be involved in the research process were present. Scientists representing different disciplines such as biophysical science, economics and social science were specifically asked to attend. We also attempted to provide an environment where everyone felt comfortable to give their views and where these views were shared. Emphasis was placed on using words like 'we', 'all' and 'together'. Emphasis was also placed on creating and maintaining an environment where farmers and scientists held positive rather than negative attitudes towards each other. During any decision making we attempted to have the group make their decisions on the data that had been presented in the open forum. We then checked to make sure that everyone was happy with the decisions made. Throughout the process we placed importance on showing the value of farmers' and scientists' knowledge.

We drew on the experience of farmers and scientists with different research. As they told their experiences, we prompted them using questions

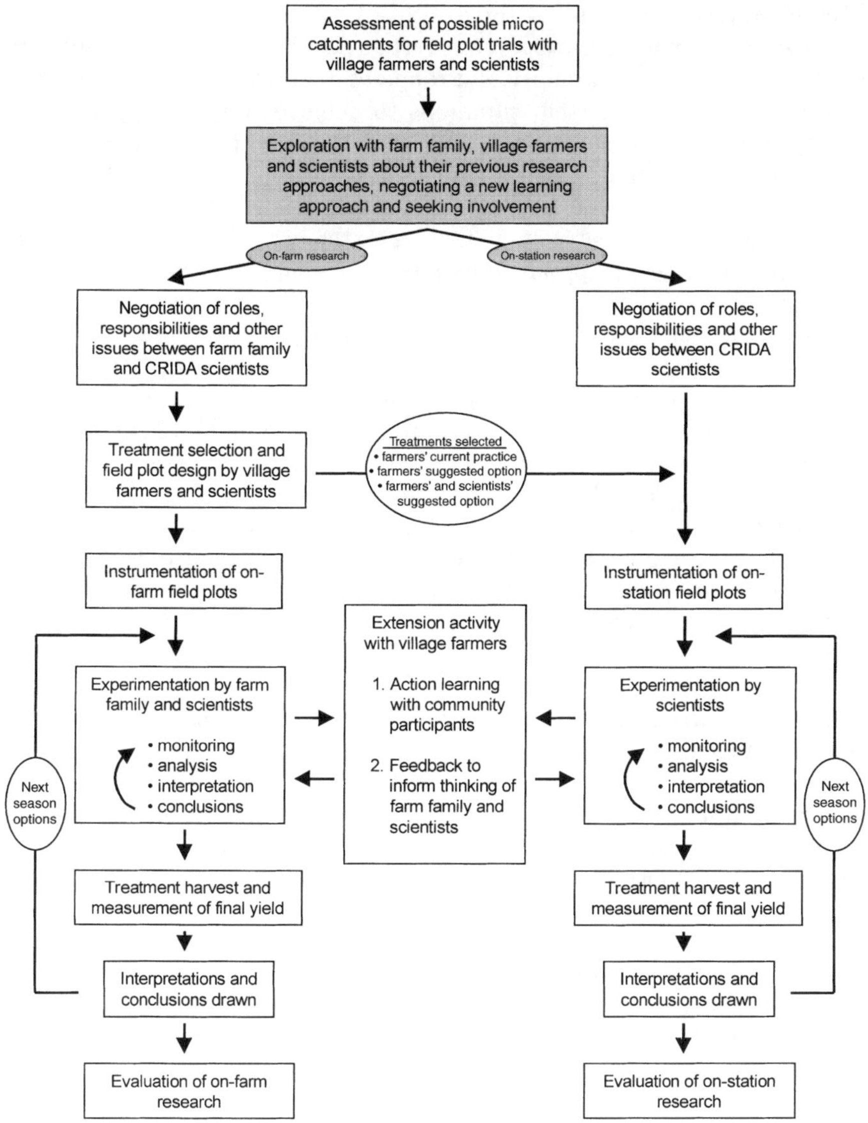

Fig. 15.2. Use of field plot trials as a tool for researching different soil management options over the cropping season in Nallavelli village.

that focused on how they carried out their research, why they used that style of research, what they were trying to achieve, and the benefits and pitfalls they had found during the research to achieve their aims.

We deconstructed some of the different research approaches mentioned by the farmers and scientists to see which elements really made the difference in achieving what they wanted in different situations. We summarized the different elements and placed them into a central pool of knowledge. We endeavoured to make this process as overt as possible to the participants. We did this by taking the summary of a range of methods and techniques (that showed usefulness in certain situations to achieve certain things) and by placing them into an imaginary space for all to share and comment on. We then spent time examining some of the underlying assumptions behind them.

Time was then spent exploring with participants what they would like to achieve if they were to carry out a 'new' research activity. We questioned the group on their issues and problems and sought to clarify a purpose and a series of objectives.

We then facilitated the group to revisit their knowledge pool to select appropriate methods and techniques to achieve the purpose and aims of their 'new' research activity. In some cases, we had to facilitate the group to think of new methods or techniques to fill certain gaps where there was no apparent appropriate method or technique available. To do this, we generated different ideas from the group and assessed them in relation to their benefits and pitfalls in achieving what was desired. We then did an overview of the 'new' research process to see whether it met the farmers' and scientists' expectations.

Farmers and scientists were then asked whether they were keen to participate in this 'new' research activity.

The Observations (observations on facilitating the 'negotiated learning and action system')

In the beginning, the suitable environment we envisaged for conducting negotiation did not exist. During the course of the discussion, the environment became more suitable. It was difficult to facilitate a discussion where everyone could have input; however, this became easier over time. Some of the farmers and scientists tried to stop others from dominating the conversation and stated that everyone should have a chance to speak. Over time, farmers and scientists that were using language that depicted an 'us and them' attitude began using words like 'we' in reference to the group in its entirety. This was also apparent between farmers and between scientists where there appeared to be power imbalances.

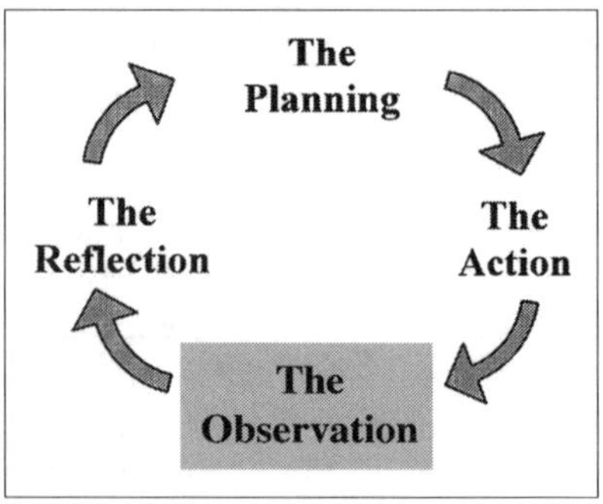

It was difficult to have farmers and scientists make decisions just on the information that had come from the group. Farmers and scientists seemed to respond well to hearing that their knowledge was of value and complementary. Some appeared to already believe this, most of the others appeared to only believe this after experiences had been shared or towards the end of the discussion and others appeared to remain sceptical. Some obvious changes in farmers' and scientists' attitudes towards each other were noted during the process.

When we assessed previously used research approaches, some of the farmers and scientists began to challenge each other and proceeded to tell each other what the other should do. Others sat back and listened. As we began prompting with questions, the challenging became less. During the questions, scientists and farmers appeared to become interested in what was being said. Farmers' approaches often constituted trial and error and were carried out on their farms. Biophysical scientists' approaches often constituted the testing of hypotheses and were carried out on research stations. Scientists from other disciplines suggested alternative research approaches that they had experienced.

There was much discussion on the benefits and pitfalls of each other's approaches. The main issues raised were in relation to accuracy and relevance. By the end of assessing previous approaches, farmers and scientists tended to hold a general view that both approaches had some value, where farmers' research took place in more relevant settings and scientists' research provided more rigorous results. As we highlighted the value of the different types of knowledge present, everyone started to nod and smile.

Sometimes it was difficult to deconstruct the research approaches and other times it was simple. At other times we chose not to deconstruct them at all. When approaches were deconstructed the pieces often represented specific methods or techniques, some of which were clearly identifiable and were the result of formal training. Others apparently had no formal label. When specific outcomes or priorities were mentioned these seemed to be quite different between the farmers and scientists. For example, farmers appeared to place priority on improving their crops and increasing the productivity of the soil and scientists appeared to place priority on achieving accurate results. When farmers and scientists expressed general outcomes they were less different. As time went by, specific and general outcomes and priorities for the 'new' research became more similar between everyone involved.

When asked whether or not they would like to be involved in the 'new' research, the farm family stated that it would be fine to carry it out on their farm and two conditions were discussed. The first was an issue of compensation and the farm family wanted to be compensated for any loss in yield compared with what they would normally obtain from what they were currently doing. Secondly, some priority issues in the village were discussed and it was felt that the project should consider these issues in the development of further research. Scientists appeared to accept these conditions.

Scientists then stated that they too would like to carry out the 'new' research with the farmers. Some scientists indicated that they would still like to carry out the 'new' research on the research stations as well. These scientists stated that it would be a backup for the on-farm research. There appeared to be some opposing views. One scientist asked 'From what we have heard, why don't we just do it all on farm?' The discussion tended to fade and it seemed that the decision to do on-station research in conjunction with on-farm research could be decided later. Farmers stated that if research was carried out on the research station then it should parallel what is done on farm. Some of the scientists and farmers suggested that the entire village should be involved in the selection of any field plot treatments.

When both farmers and scientists stated that they would like to be involved in the 'new' research there appeared to be sighs of relief by both farmers and scientists. One farmer stated that 'if we do it together it will work. Otherwise it won't.' Everyone showed signs of agreement with excitement and enthusiasm to continue.

After the event, a group of farmers approached one of the facilitators and expressed that they were happy with what went on. A representative for the group stated, 'We would like to thank you. This is the first time that anyone has told us that our knowledge was of value.'

Another farmer approached later and declared, 'You have done a good thing here. Now we can try some things that we would have liked to have tried a long time ago. It was always too risky before.'

The Reflection (reflectively evaluating the results from facilitating the 'negotiated learning and action system')

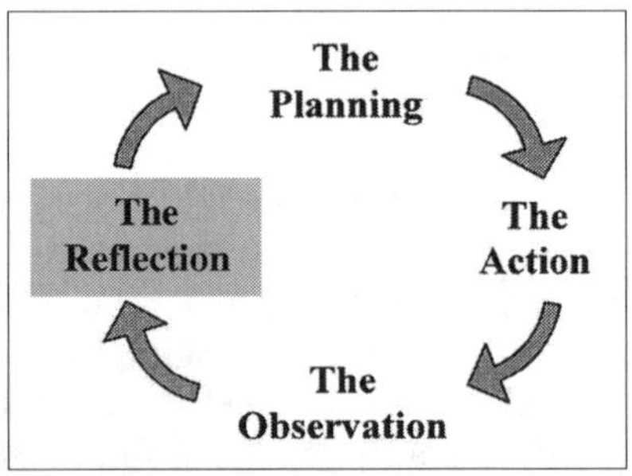

We can suggest why a suitable environment for facilitating a 'negotiated learning and action system' was not present when we began working with farmers and scientists on this project. This relates to the perceptions of farmers and scientists about each other, particularly in relation to conducting activities. These perceptions are based on an individual's own experiences or on the passing on of others' experiences.

Through history, farmers' knowledge has often been ignored during research and other developmental initiatives. Using Pretty's typology (1995) to articulate participation in the past, participation has predominantly been manipulative, passive or consultative, sometimes functional or for material incentive and seldom interactive or self-mobilizing. For these reasons, beneficiaries of research have a long history of receiving less from the research process than the organizations or people that initiate or implement them. This has often left farmers with negative experiences of working with these other institutions, leading to mistrust and uncertainty. Having infrequently worked with farmers before, scientists also feel uncertain and apprehensive, particularly with respect to control over the research agenda. Much of traditional science had been based on providing controlled environments.

Building an environment of trust and certainty

TO OVERCOME A LACK OF TRUST, WE PLACED EMPHASIS ON FOSTERING AN ENVIRONMENT OF LISTENING RATHER THAN TELLING This involved farmers and scientists fostering a new way of communicating with each other. Scientists had to be held back from telling farmers what to do and farmers had to reduce their expectations of being told the answers. Focusing this part of the process on sharing experiences and constructive inquiry worked well. It was also necessary to try to ensure input from a cross-section of participants. To do this, the focus of selecting participants to share their experiences was based on their ability to contribute to the sharing of experiences and not their position of power. This was critical as positions of power not only occurred between farmers and scientists, but also between scientists and between farmers.

TO OVERCOME UNCERTAINTY, WE USED PURPOSEFUL QUESTIONING TO LINK EACH PARTICIPANT'S OWN KNOWLEDGE WITH THEIR PAST APPROACHES OF INVESTIGATION OR RESEARCH In some instances, these approaches were deconstructed into smaller pieces that represented methods and techniques. This purposeful questioning led to different perceptions being brought into the open forum and enabled everyone present to judge past approaches, methods and techniques in relation to achieving the purpose of the research. Through this process, participants themselves were given the opportunity to discover the value of these different methods in certain situations.

The rationale behind this purposeful questioning was to separate people from previous problems in achieving the purpose. This was more beneficial than relating methods back to individuals or groups of individuals, i.e. farmers tend to do things one way, scientists another, which tends to foster the 'them and us' attitude. A second rationale for purposeful questioning was to draw out key principles from the group to be used to produce some common criteria on which to later construct a 'new' process. This involved exploring some of the underlying assumptions behind approaches, methods and techniques.

OWNERSHIP AND COMMITMENT WERE FOSTERED BY CONSTRUCTING A 'NEW' PROCESS USING THE APPROACHES, METHODS AND TECHNIQUES THAT THE ENTIRE GROUP FOUND BENEFICIAL IN CERTAIN SITUATIONS This worked well because everyone could see where their knowledge was contributing to the 'new' approach. The result was a common framework built by both farmers and scientists. The process of construction also highlighted an interdependency of participants in the group in relation to achieving the 'new' approach. In the past, scientists have worked independently of farmers (and vice versa) and farmers have been dependent on scientists to produce technologies.

Language proved to be a good indicator of change in perceptions of the value of each other's knowledge. For example, some participants who initially challenged other individuals began to challenge approaches instead. A shift was noted from challenging approaches to providing constructive criticism. We found that in many cases we had to cut short the time allowed for the process of deconstructing and reconstructing. For many of us, this was new and we frequently underestimated the time it would take. This not only affected the process, but also left less time to do follow-up processes.

Incorporating systems thinking

How to carry out multidisciplinary projects has been an unforeseen challenge. Problems with conducting projects of this type exist because of the different perspectives and approaches within single disciplines. Patton (1990) describes how different disciplines have evolved over time by focusing on different core questions. This has resulted in theoretical traditions and orientation that have produced implications for how researchers operate. In essence, researchers within single disciplines now come together to plan, implement and evaluate activities in many different ways, and these different ways are often conflicting and competitive. In addition, researchers within specific disciplines are likely to view the world revolving around their own discipline. That is, their discipline may be seen as a focal point to which the other disciplines link.

HAVING SCIENTISTS FROM DIFFERENT DISCIPLINES PLANNING, THINKING AND APPLYING RESEARCH IN CONJUNCTION WITH FARMERS ENABLED A MORE SYSTEMIC APPROACH TO EXPLORING THE COMPLEXITY OF THE SYSTEM Farmers, who work and live in these systems and learn to manage these changing environments through necessity, could be challenged to explore certain parts of their system in more detail. Scientists were challenged to apply their disciplinary concepts in a systems framework, so that they could make connections between their own and other disciplines, and to explore the interrelationships between these.

Not only does a multidisciplinary approach need to overcome problems of researchers' different roles and ways of research, but it also implies an understanding of the links between the disciplines. For example, a project containing social, economic and biophysical elements requires researchers to investigate and understand the interrelationships and linkages between all three. This is a task rarely done in applied research.

IT SEEMED TO BE CRUCIAL FOR FARMERS TO BE INVOLVED WITH SCIENTISTS AT THE VERY START OF RESEARCH PROJECTS This was particularly important at the stage where decisions were made on whether to conduct research, well before the planning stages. Each party had to come to conclusions themselves that they each had valuable knowledge to offer in the research process. Emphasis was placed on this because, when these values are recognized before planning occurs, then the planning (and subsequently the rest of the research process) can embed a methodology that represents more interdisciplinary and relevant ways of working.

If a technology is developed through the single discipline-based methodology and is simply put through a testing phase to see whether it is applicable with respect to other areas, then, if it is not applicable, these scientists go back to the drawing board and possibly repeat the same cycle. This is not so much different from traditional approaches that left out farmers and scientists from different disciplines in the entire research process.

The question today is not whether multidisciplinary research, development and extension should be carried out, but, rather, how to go about operating and making decisions in an effective multidisciplinary way. This entails drawing on the information, approaches and concepts of the single disciplines, recognizing them for strengths and weaknesses to address the task, and then integrating or modifying them in such a way as to improve the effectiveness and efficiency of a project. Facilitating questioning to draw out these differences enhanced this further.

IT SEEMED IMPORTANT FOR FARMERS AND SCIENTISTS FROM DIFFERENT DISCIPLINES TO INFORM EACH OTHER OF THEIR WAYS OF RESEARCH Doing this through the sharing of experiences was an effective means, because others have a context and basis to decide for themselves the value in these different methods. If each party presented their different way of working in isolation from the context and the purpose, others would find it difficult to judge these methods. Sharing of experiences allowed for all to form trust in both methodology and the people. Everyone could make a decision to agree to be involved based on new knowledge of the benefits of different ways of working. In any case, the decision to be involved can feel less risky. Each party is reassured and, in the above case, some became convinced that the research to achieve their purpose would not work unless everyone did contribute.

Informing the decision of whether to participate or not

Much has been documented about the transfer of technology model of extension, illustrating that it had not provided an equitable or effective means of technology innovation. The transfer of technology model presents extension as a one-way flow of information from scientists to farmers where scientists also controlled the planning, implementation and dissemination of research and became custodians of knowledge. This model implied that research was something that was done for farmers and farmers' land rather than with farmers.

WE INITIATED NEGOTIATION (AS A FORM OF PARTICIPATION) TO REPRESENT OUR BELIEF THAT FARMERS HAD A RIGHT TO DECIDE WHETHER OR NOT THEY WANTED TO BE PART OF THE RESEARCH PROCESS When we made this decision we were quite conscious that we did not want to force people to participate. The recognition that participatory processes have a number of advantages over previous approaches has resulted in the production of many generic participatory processes and many practitioners being told (by funding bodies, managers, research organizations, etc.) that they 'must do' participatory research. A paradox has developed where we have moved from one transfer paradigm to another. That is, transfer of technology has become transfer of methodology and the reason we moved away from transfer of technology to an alternative approach now underlies this alternative approach. In order not to comply with a 'transfer of methodology' approach and enforce participation, we chose negotiated learning as a way of articulating participation.

In this example, farmers may not have agreed to be involved in the research process and scientists would have been left to carry out their research on stations. Scientists would not have the opportunity to do their research in the more complex environment needed to develop innovations relevant to farmers' needs, particularly in rainfed areas. Scientists may not have agreed to work with farmers and farmers would have missed the opportunity to have input into research priorities, gain from scientists' knowledge of producing accurate data, and have access to resources to enable them to explore alternatives without high risk to their livelihoods. Scientists from different disciplines may not have agreed to work together either, missing the opportunities to explore the interrelationships between biophysical, economic and social phenomena.

FACILITATING NEGOTIATED LEARNING MEANT THAT, THROUGHOUT THE RESEARCH PROCESS, THE PRINCIPLES, PURPOSE, PROCESS AND PEOPLE WERE UP FOR NEGOTIATION WITH THOSE INVOLVED AT ANY POINT IN TIME We found this particularly difficult because it meant that we, as facilitators, had to forgo our individual needs and preferences for those suggested by the group, even if we thought we had the best method, idea or way of achieving something. The facilitator's role was to bring everyone along in the process together and this required everyone to be part of the decision making to have ownership and commitment to it. The notion behind this was that, in the long run, the group moving along together, learning together and making decisions together will prove to be more beneficial than if one or two people chose the direction of the group or made some of the fundamental decisions about the group's direction. One of the most important points here was that the facilitator had to have trust in the process because, in the short term, the process tended to take longer and be more complex.

Facilitating a 'negotiated learning and action system' meant that, although we came up with what we thought were the most appropriate methods for the given situation, these could be deleted, replaced or modified by others in the process. Any changes required that we would have to accept and design new processes to incorporate them. This necessitated that specific processes were in place that stimulated periodic critical inquiry about the current and future process. The underlying assumption behind negotiated learning is that the process is guided by all those involved in the process rather than the facilitator. This meant that we had to create an environment in which all those in the process had an opportunity to safely question the process itself.

Another technique that was useful was to explore with participants (at every opportunity) the variety of options available for the 'new' research. Purposeful inquiry into the benefits and pitfalls of different research methodologies aided in this examination of different approaches. When problems of different approaches were identified, we also looked at why these problems may have occurred and ways of overcoming them.

WE FOUND THAT IT WAS USEFUL TO INFORM POTENTIAL PARTICIPANTS ABOUT A VARIETY OF OPTIONS AND POSSIBILITIES BEFORE THEY DECIDED TO PARTICIPATE In this way, participants could make a judgement on their level, if any, of participation based on a facilitated process. The aim was to present the benefits and possibilities of a whole range of approaches rather than promoting one particular process. The emphasis was to provide skills of critically thinking about this range of approaches and being able to select the most appropriate one for a given task. They could also make a decision on whether they felt their learning objectives were going to be met and whether they thought others were committed. That is, the process also intended to enlighten both farmers and scientists about each other's motivations, principles, purpose and methods so that everyone could make a more informed decision on whether or not to be involved.

Empirical Line 2: Using the Portable Rainfall Simulator as a Tool for Researching the Effect of Rainfall Intensity on Different Soil Management Practices (see Fig. 15.3)

The Action (facilitating the 'negotiated learning and action system')

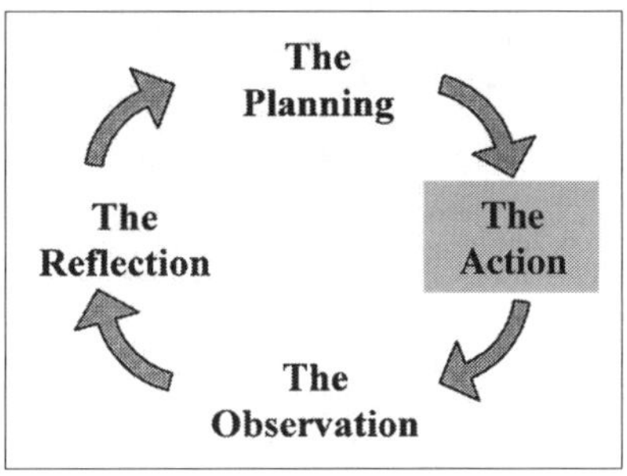

We used a rainfall simulator to provide a simulated experience of rainfall on a variety of soil surface management practices. First, a farm family was contacted about whether we could run a field day activity in their field. If this was acceptable, we contacted village farmers to inform them of the field day. We encouraged scientists from a range of disciplines to be involved in this research activity with farmers. On the day, we organized the construction of the rainfall simulator by scientists, the farmer and other available village farmers. We then set the scene by informally introducing people and briefly explaining the use of the simulator. Once the simulator was ready, we facilitated farmers and scientists in the planning of an activity using the rainfall simulator. They planned what they were going to do and what they would monitor along the way. We facilitated farmers and scientists in negotiations of possible treatments to investigate. We then encouraged farmers and scientists to implement the treatments together.

A simulation was conducted. Purposeful questions were raised periodically. During the planning stage, questions were focused around the purpose of the activity, with the facilitators often iterating between the purpose and

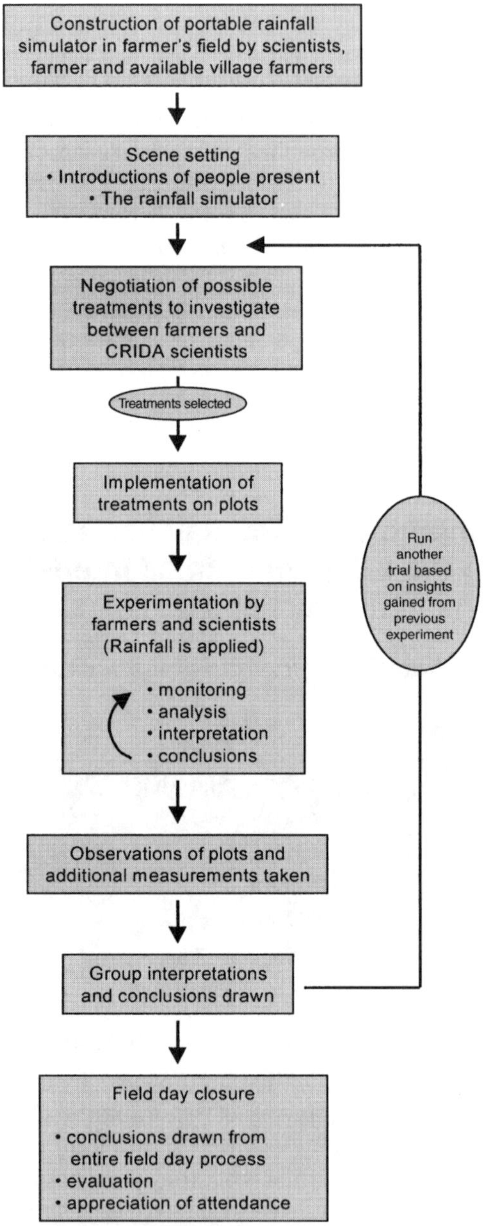

Fig. 15.3. Use of a portable rainfall simulator in Nallavelli village as a tool for researching the effect of rainfall intensity on different soil management options.

process columns. During the implementation participants were allowed to interact with the system; we then explored this interaction. Questions were asked of both farmers and scientists about why they chose to implement and monitor the system the way they did. During the action stage, we questioned them about the biophysical interactions that were being observed.

Through the use of questions the participants were guided through a process of reflection and drawing conclusions, after which they were assisted to explore what they would do differently if they were able to select two more treatments and run the activity. The conclusions drawn from the first treatments were then used to modify the original treatment plan and they then implemented their two 'new' treatments and the cycle continued.

The general steps followed during the rainfall simulator activities were:

- set the theme for the day – exploring soil and water process responses to different management options;
- participants nominate treatments to investigate (plan);
- participants construct the treatments (act);
- rainfall is applied (act);
- monitoring and discussion of the effects by participants (observe);
- small group discussions are held (reflect); and
- decide on further treatments as a result of outcomes and learnings from the initial rainfall simulator runs (another learning cycle).

The Observations (observations on facilitating the 'negotiated learning and action system')

The rainfall simulator (RFS) demonstrated water infiltration and runoff from the plots. Farmers were enthusiastic in helping construct the RFS as well as the implementation of treatments. Farmers seemed to appreciate being asked to put their own treatments up for testing. We went through a number of research cycles on the day. Farmers and scientists appeared to become more relaxed with each cycle. Initially, some scientists tried to tell farmers what they should be using as treatments. Other scientists told them to be quiet and just watch and let the farmers select their own treatments.

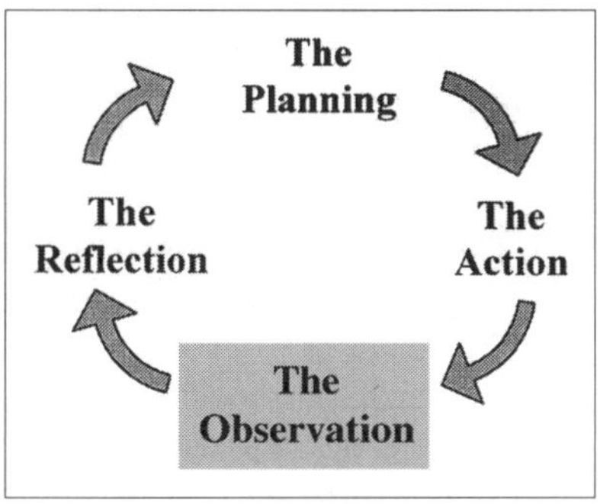

At times the team found themselves wanting to revert to a default extension process by using the RFS as a demonstration tool to 'teach' farmers improved fallow management techniques and reinforce scientific concepts that 'should be adopted'. During these times the facilitator had to encourage a 'have faith and trust' concept.

During the course of the field day, some scientists started taking on the role of facilitator. They asked different questions of farmers in relation to their own disciplines. The RFS appeared to show some degree of novelty and farmers seemed keen to be involved. The water and silt levels in collection tanks, the influence of rainfall on the surface and the infiltration and runoff rates were all observed and reflected upon individually, between one another and then in a group. Scientists looked as if they were happy with the data they were collecting. Farmers claimed that they found it valuable to vary rainfall intensity during the activity to more closely replicate rainfall patterns. Farmers and scientists came up with some further treatments that they wanted to test, based on what they had just seen.

Initially, the treatments chosen appeared to be treatments of which farmers and scientists were sure. Most expressed that they knew 'the answer'. In the following test, both farmers and scientists appeared anxious to see what was going to happen. The equipment created lively discussion on management practices to control soil loss. Farmers and scientists measured infiltration by using a metal probe or digging with a shovel/mattock with no prompting. They also measured runoff by comparing the amount and level of water in the vacuum drums. Farmers used the colour of water in the vacuum drums as an indicator of sediment and soil loss.

The RFS attracted both women and men farmers, as well as children in the village. The participants liked the opportunity to participate, test their ideas and see effects of traditional practices. When tribal women farmers were present, they were extremely active in selecting and modifying treatments and in providing good debate. The selection and application of treatments always inspired long and willing debate. Farmers and scientists both appeared to learn from being able to visualize raindrop impact on the soil. The RFS as a tool was successful in creating thought-provoking discussion on management practices to control soil loss.

The Reflections (reflectively evaluating the results from facilitating the 'negotiated learning and action system')

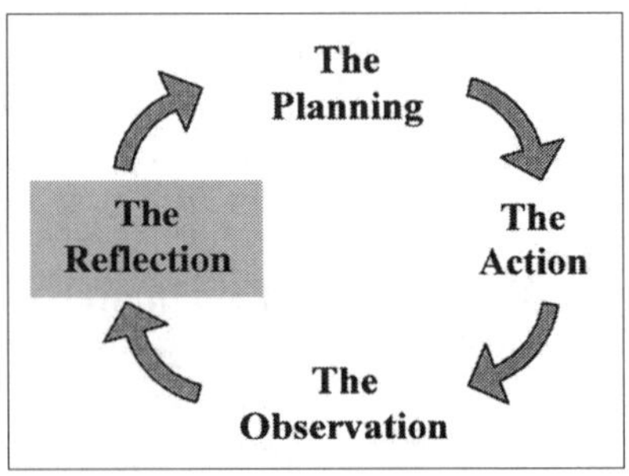

This example provided some additional reflections on the facilitation of the 'negotiated learning and action system'. Although many of the reflections above also apply to this example, what is presented here are some of the additional points of interest.

The tool generated enthusiasm and energy because it operated in the real environment and illustrated basic principles that were relevant and useful. The tool also provided rapid cycling (action research cycling) and a fast turnover of cause and effect. For these reasons, the tool was both effective and efficient.

As mentioned in the previous example, farmers and scientists often enter into situations with each other that are not trusting. The tool enabled farmers and scientists to initially test the tool rather than each other. Once the tool passed the test of credibility, further time enabled each to build up a trusting relationship with the other. Rapid cycling enhanced this. It was important for the facilitators to have a strong vision of the process so that we could facilitate in such a way as to manage any individual domination during the process. We encouraged a present view of exploring what will happen rather than a knowing of what will happen.

We also emphasized two matters of importance. First, that the simulator was only creating a common phenomenon (i.e. rainfall) and, second, that the simulated experience was not the real world but a way of exploring and postulating about it. It is the process that tended to be important, rather than the content that came out of each run. The simulation was a replicate of the world as best we could achieve, and provided a forum where all participants could investigate innovations.

The simulated experience not only allowed for biophysical components to be studied using a hard systems approach, but also enabled people to interact with the system and interactions could be explored through critical questioning, within a soft systems perspective. Farmer learning was enhanced when this questioning was carried out by scientists from a range of disciplines or by farmers that carried out other types of tasks (e.g. role differences between genders). Scientist learning was enhanced when questioning was performed by farmers or by scientists from different disciplines.

The repeated use of the RFS aided the development of skills as some participants started to take on the facilitation themselves. We believe this occurred because participants saw the benefits of an inquiry approach for discovering aspects with regard to the learning objective and wanted others to discover this also. This effect occurred more and more as the RFS was run over time and reflected the change in approach to research within the project over time. Often, the same scientists were involved in running the simulator with different groups of farmers. The fact that scientists could move outside their comfort zone and take on a role of facilitation illustrated the safe environment the tool tended to generate. Organizational support (at the highest local level) for scientists to do different things seemed to contribute to this. On reflection, the RFS could also offer a mechanism to train in different ways of facilitation.

The process of critical inquiry and reflection through the RFS activity was the foundation of the research. This was because the questioning, when focused around purpose, could provide relevant and appropriate interpretations and conclusions about the results. Action research created an experience involving all participants, which was then a focal point for reflective inquiry where generalizations and conclusions could be drawn to inform the process of negotiation for the selection of future treatments. Ellinor and Gerard (1998) suggest that inquiry is dependent first on intention, second on the type and quality of question, and last on quantity. This paralleled our thinking when constructing questions so that they focused primarily on the question's purpose and secondly on quality (as we knew we could phrase the question in a variety of ways) and finally on quantity (and, in most cases, the quantity of questions was not a consideration). 'Inquiry is about asking questions and holding an attitude of curiosity, opening the door for new insights... Reflection is about holding the door open long enough for new perceptions to emerge' (Ellinor and Gerard, 1998).

Conclusions

The 'negotiated learning and action system' that resulted from the use of the '5P framework' was a system of inquiry that led to community, team and organizational change. Its key principles of systems thinking, negotiated learning and action research not only complemented one another but also provided the learning system with its own synergistic qualities. The information that was constructed and generated by this system was of greater relevance and more equally distributed between and representative of those within the system, compared with that from traditional approaches. Inquiry within the system was purposeful and critical and provided the system's primary foundation. The system was dynamic, emergent and continuously improving.

The authors suggest that this research process has applicability in a wide variety of research situations that aim to improve the sustainability of complex environments. With this in mind, practitioners are required to have trust in the system and its underlying philosophies, principles and methods. The facilitator must place priority on the preferences and judgements made by those within the system, over their own preferences and judgements, in an effort to have all participants move forward together in the system of inquiry and to direct their own destiny. The '5P framework' provides the facilitator (and those involved in the system) with a mechanism to articulate and explore the implications of changes that occur in relation to the system's purpose, principles, paradigm, people and processes. Importantly, the facilitator needs to aid the development of trust and certainty between participants while engendering ownership and commitment. These are essential criteria for effective system function.

Acknowledgements

Thanks go to the people of Nallavelli, Pampanur and Madabhavi villages for their involvement, openness and sharing of ideas. Extension agents, research associates, technical staff and scientists from both the Indian and Australian teams are acknowledged for being prepared to be challenged by an unfamiliar system of research. CRIDA, DNR and DPI have provided the staff, resources and support to undertake this novel research. ICAR and ACIAR have initiated and funded this work. Special thanks to Deborah Maxwell for editing and supporting the compilation of this chapter.

References

Dick, B. (1993) *You Want to do an Action Research Thesis? How to Conduct and Report Action Research.* Interchange, Brisbane.

Ellinor, L. and Gerard, G. (1998) *Dialogue: Creating and Sustaining Collaborative Partnerships at Work – Rediscover the Transforming Power of Conversation.* John Wiley, New York.

King, C.A., Jiggins, J. and Coutts, J. (1998) Unearthing paradox. In: *Proceedings of the Third European Farming Systems Research and Extension Symposium*, November–December, Pretoria, South Africa.

Kolb, D.A. (1984) *Experiential Learning: Experience as the Source of Learning and Development.* Prentice-Hall, Englewood Cliffs, New Jersey.

Krueger, R.A. (1988) *Focus Groups: a Practical Guide for Applied Research.* Sage Publications, London.

Patton, M.Q. (1990) *Qualitative Evaluation and Research Methods* (2nd edn). Sage Publications, London, pp. 64–91.

Pretty, J.N. (1995) Participatory learning for sustainable agriculture. *World Development* 23(8), 1247–1263.

Progress Report (1997) *Tools and Indicators for Planning Sustainable Soil Management on Semi-Arid Farms and Watersheds.* ACIAR Project 9435. Central Research Institute for Dryland Agriculture, Santoshnagar, Hyderbad, Andrah Pradesh, March–October, 1997.

Roling, N.G. and Jiggins, J. (1998) The ecological knowledge system. In: Roling, N.G. and Wagemakers, M.A.E. (eds) *Facilitating Sustainable Agriculture: Participatory Learning and Adaptive Management in Times of Environmental Uncertainty.* Cambridge University Press, Cambridge, UK, pp. 283–311.

Stewart, D.W. and Shamdasani, P.N. (1990) *Focus Groups: Theory and Practice.* Applied Social Research Methods Series, Vol. 20. Sage Publications, London.

Zuber-Skerrit, O. (1995) Models for action research in postgraduate education. Paper presented at the 1995 Action Learning and Action Research Conference, The University of Queensland, Australia.

Research Approaches to Developing Sustainable Management Practices on Australian Vertisols

16

R. CONNOLLY,[1] N. DALGLEISH,[2] K. COUGHLAN,[3] D.M. FREEBAIRN[1] AND M.E. PROBERT[2]

[1]*Queensland Department of Natural Resources, Toowoomba, Queensland, Australia;* [2]*CSIRO, Tropical Agriculture, Indooroopilly, Brisbane, Queensland, Australia;* [3]*ACIAR (Australian Centre for International Agricultural Research), Canberra, Australian Capital Territory, Australia*

Key words: Australia, Vertisols, agricultural research, soil management, soil erosion, soil fertility, environmental impact, yields, *Zea mays*, maize

Introduction

Australia has 80 million ha of Vertisols, the largest area of Vertisols in a single country in the world (Ahmad, 1996; Fig. 16.1). In arid and semi-arid areas, native pastures on Vertisols are grazed by sheep and cattle. In higher rainfall areas, improved pasture is grazed. In the east and south of Australia a range of winter (wheat, safflower, barley and oats) and summer (sorghum, maize, soybean, cotton, sunflower and millet) crops are grown under natural rainfall. Cotton, grain, fodder crops and rice are also grown with irrigation. Cropping is highly mechanized, and large areas are managed using high horsepower and heavy equipment. For example, a 2000-ha cropping programme may use a 336 kW (450 hp) tractor weighing 20 t, pulling 20-m-wide equipment.

In this chapter, we summarize issues that are important for the long-term, sustainable management of Vertisols in Australia. Research that aims to address these issues is discussed. One research approach is to use simulation

(eds J.K. Syers, F.W.T. Penning de Vries and P. Nyamudeza)

models. An example of how models can be used to add value to a land-shaping experiment at Chisumbanje, Zimbabwe, is given.

Land Use Related Problems

Vertisols are susceptible to a number of land use related problems. Most important are:

1. *Erosion*: because of their self-mulching nature and low aggregate density, Vertisols are highly erodible (Freebairn *et al.*, 1996b). Vertisols are sometimes cropped on land slopes as steep as 10%, but more typically 1–3%. Sloping land combined with high-intensity rainfall and bare soil, especially in the northern half of Australia, can lead to severe erosion. Soil conservation works, particularly contour banks, have reduced erosion by reducing slope length. Adoption of cropping practices that leave the soil covered (zero and minimum tillage) has also reduced the incidence of severe erosion events. However,

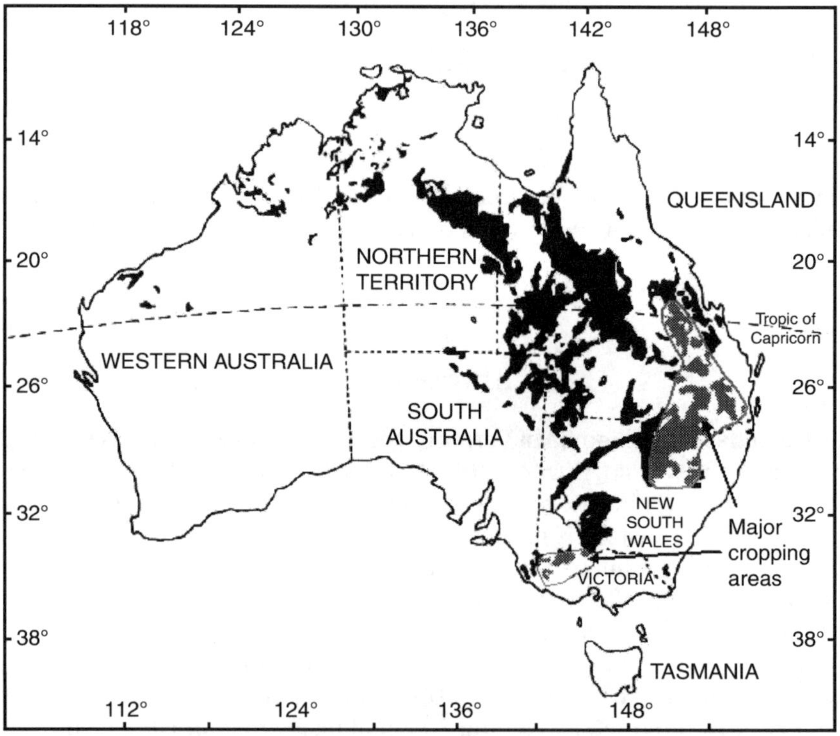

Fig. 16.1. Location of Vertisols and major areas of cropped Vertisols in Australia (from Ahmad, 1996).

poor maintenance or improper installation of contour banks and agronomic problems associated with retaining stubble mean erosion is still an important issue for the sustainable management of cropping lands.

2. *Rundown in soil physical and chemical properties*: tillage and cropping have led to the rundown of a range of soil physical and chemical properties. Disturbance associated with tillage and changed organic matter cycling has caused organic matter levels in the soil to decline, often to half their initial levels (Dalal and Mayer, 1986b). Soil fertility declines as organic matter runs down. Lower organic matter levels combined with tillage disturbance have led to reduced aggregate stability of surface soil and increased surface crusting, causing restricted infiltration and problems with waterlogging, seedbed preparation and efficient use of rainfall by crops (Freebairn *et al.*, 1986b). Machinery wheel tracks and smearing associated with tillage operations cause compaction of sub-surface soil layers. Compacted soils are less permeable, have changed water-holding properties and have increased resistance to penetration by roots (Gupta *et al.*, 1989).

3. *Rainfall use efficiency and cropping management in an uncertain climate*: water stress is a common limitation to crop production on Australian Vertisols (Freebairn *et al.*, 1986a). Rainfall in Australia is highly variable and potential evaporation is much greater than rainfall. In the cropped areas, median rainfall varies from about 550 to 700 mm year^{-1} and the coefficient of variation of annual rainfall is typically greater than 25%. Average annual evaporation is typically 2000 mm. Unlike other countries with Vertisols, many areas in Australia have no well-defined 'growing season' and crops are grown partly on in-crop rainfall and partly on water stored in the soil at planting. As a result, the soil is often left uncropped (in fallow) for up to 12 months to accumulate soil water. But the efficiency of conversion of rainfall over fallows into crop-usable soil water is low, varying from 0 to 35% (Freebairn *et al.*, 1986a). Much of the rain that falls over fallows is lost to evaporation because rainfall events are typically about the same size as the evaporation deficit (difference between air-dry soil and crop lower limit in the top 100 mm, from which rapid evaporation occurs) (Thorburn *et al.*, 1989).

4. *Impacts on the environment*: more recently, concern has increased over the impact of cropping on the off-farm environment. Movement of sediment, pesticides and nutrients in runoff water, drift of pesticide during spraying operations and leaching of pesticides and nutrients into groundwater have the potential to cause damage to the environment. Changed land use also affects catchment hydrology, leading to salinity outbreaks downslope as a result of rising saline water tables and to increased salinity in streams and water bodies (Viessman *et al.*, 1977).

5. *Grazing lands*: overgrazing has led to erosion, degradation of native pastures and the emergence of woody weeds. Management is made more difficult because of highly variable rainfall, declining real commodity prices and sometimes the small size of holdings.

With increasing costs of production, variable commodity prices, rundown in soil properties and environmental pressures, farmers in Australia are constantly pressed to evaluate and improve their management systems. This often involves moving toward management systems that include opportunity cropping, minimum tillage, strategic crop rotations and use of legumes and pasture leys (Wylie, 1999). Rummery and Coleman (1999), for example, show minimum tillage systems to be consistently more profitable than systems where intense tillage is used. Awareness of environmental issues among farmers and land managers is high. As an indication of the level of interest, participation in 'Landcare' groups (small community groups working towards addressing environmental and land use problems) has grown exponentially since the Landcare movement started in 1985 (Campbell, 1992).

Research Approaches

Erosion

Research into erosion processes has been very successful in developing systems to minimize erosion risk while maintaining crop and farm productivity. The most important progress has been made with:

1. *Soil conservation structures*: contour banks (bunds) spaced between 30 and 120 m apart (spacing is determined based on land slope, soil type and rainfall characteristics) are a very important way of controlling erosion on sloping country (QDPI, 1966). The contour banks intersect runoff, reducing slope length and minimizing the formation of rills or gullies. Water is transported at lower velocity off the paddock to grassed waterways. On nearly flat areas, growing crops in strips at right angles to flow (strip cropping) has proved successful in preventing erosion damage from flood flow. Lately interest has also been high in 'controlled traffic', where machinery follows defined tracks, rather than driving all over the paddock, and row furrows help to control movement of runoff (Fig. 16.2).
2. *Cover*: the importance of maintaining cover on the soil surface to reduce raindrop impact and increase flow tortuosity has been known for some time (e.g. Fig. 16.3), and is a principle well adopted by the farming community. As a result, zero and minimum tillage systems, which aim to maintain cover levels and reduce soil disturbance, have become increasingly common.
3. *Reducing runoff*: runoff from Vertisols is related to both soil water content and surface conditions, particularly presence of a surface crust (Freebairn *et al.*, 1986a). Management that keeps the soil dry, e.g. with increased crop frequency, and maintains surface cover leads to reduced runoff, hence reduced erosion potential.
4. *Simulation of effects of erosion on productivity*: computer models have been used to simulate long-term soil loss from a soil profile, effects on crop

Fig. 16.2. Tracks in a paddock farmed with controlled traffic. Machinery operators follow the line in the middle for the first year or two until the wheel tracks on each side of the rip mark become established. Machinery travels only on these tracks to improve operating efficiency, reduce erosion and reduce soil compaction.

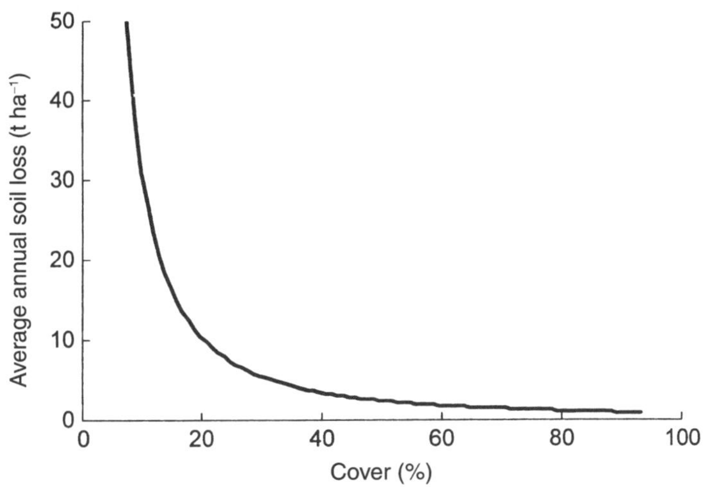

Fig. 16.3. Average annual soil loss from a Vertisol in contour bays on the eastern Darling Downs (Greenmount) for various management systems that maintain differing amounts of surface cover (Freebairn *et al.*, 1996b).

productivity and relative effectiveness of alternate management practices in reducing soil loss. Models used to simulate Australian Vertisols vary from simple decision support systems (USLE, Wischmier and Smith, 1978), through process models such as CREAMS (Knisel, 1980) and WEPP (Lane and Nearing, 1989) to soil–crop models that simulate relationships between erosion, management and productivity (PERFECT, Littleboy *et al.*, 1989; EPIC, Williams *et al.*, 1990). Figure 16.4 shows a typical application of an erosion model: the predicted effect of several fallow management strategies on erosion from a Vertisol on the eastern Darling Downs (Greenmount).

Rundown in soil properties

Declining soil fertility

When first brought into agricultural production, Vertisols were generally highly fertile. However, cropping without adequate replacement of the nutrients removed in produce ('nutrient mining') or otherwise lost from the system (e.g. as a result of erosion or burning of crop residues) has caused fertility of Vertisols to decline markedly. Levels of all nutrients have been run down, but nitrogen is commonly deficient for cereal crops. Dalal and Mayer (1986a) found uptake of total N by winter cereals on Waco soils in southern Queensland declined from 120 kg ha^{-1} to 40 kg ha^{-1} after 70 years of cropping. When nitrogen is limiting, the protein content of grain is lowered before yield is reduced. Protein in winter cereal grain grown in sub-tropical Queensland has been observed to decline from about 16% to about 8% after 50 years of cropping (Dalal *et al.*, 1991).

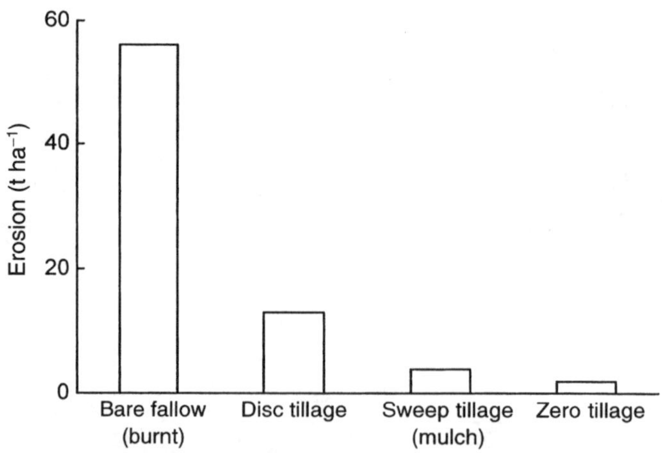

Fig. 16.4. Average annual erosion for fallow management practices on a Vertisol on the eastern Darling Downs, Queensland (Freebairn *et al.*, 1991). Results are simulated with the PERFECT model for 1912–1985, crop is winter wheat, summer fallow.

Decrease in the nitrogen supplying capability of soil is associated closely with decrease in soil organic matter content (Fig. 16.5). After cropping for between 40 and 70 years, organic carbon of surface soil layers has typically halved. The amount of organic matter in soil depends on the rate of addition of new organic matter from crop residues and the rate of decomposition of the organic matter. Under cropping, inputs have been insufficient to maintain the initial high organic matter content of the soils, while rates of decomposition are higher when the soil is disturbed by cultivation. With time, though, inputs tend to balance outputs, and the organic matter levels stabilize.

A major challenge facing the industry is how to improve organic matter inputs in current farming systems on an economically sustainable basis.

There has been considerable research into management practices to address the inadequate supply of nitrogen from cropping lands. Comprehensive fertilizer trials have been undertaken to characterize optimal fertilizer composition and rates (e.g. Ockerby *et al.*, 1993; Strong and Cooper, 1995). Strong and Cooper (1995) found that consistent application of nitrogen fertilizer at rates up to 75 kg ha^{-1} improved financial returns due to increases in grain protein concentration, but that yield response was variable because of variable rainfall. Despite this favourable result, fertilizer application has not been adopted widely by farmers, particularly in drier areas. In dry years there may be little or even a negative response to fertilizer, whereas in years when there is good in-crop rain yield potential increases markedly. Since the yield potential cannot be known at sowing, recommendations for fertilizer use in this environment are problematic.

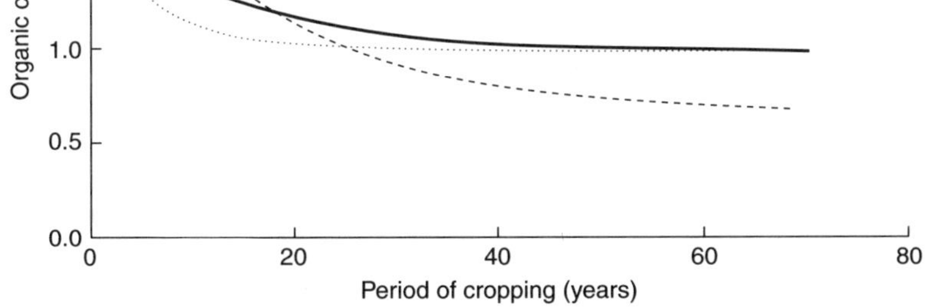

Fig. 16.5. Rundown in organic carbon with period of cropping in the 0–0.1 m deep soil layer of some Vertisols in southeast Queensland (Dalal and Mayer, 1986b). Organic carbon has run down because of disturbance due to tillage and a changed organic matter cycle.

Rotations with pasture ley and annual legume crops have shown potential, raising soil organic matter and nitrogen content (e.g. Holford and Crocker, 1997; Dalal *et al.*, 1998). Dalal *et al.* (1995) found that, after 4 years' grass + legume leys, organic carbon in a Vertisol at Warra, Queensland, increased by 20% (650 kg C ha^{-1} $year^{-1}$) relative to continuous intense tillage. Soil total N increased by 22%. Currently research is under way to evaluate the role for ley rotations to maintain/improve soil organic matter and improve nitrogen supply to crops compared with applying inorganic fertilizer. Holford (1980) found that a 3.5-year lucerne ley eliminated the need for nitrogen fertilizer for the following five wheat crops. But, in the present economic climate, pasture leys (particularly grazed leys) may have lower gross margins than cropping, occupying land and using water that could have been more profitably used by crops.

Computer models have considerable utility for extending experimental results to a wider range of soils/climates/cropping regimes and analysing the pros and cons of the various options to manage soil fertility (e.g. Probert *et al.*, 1998; Turpin *et al.*, 1998). In the FARMSCAPE programme, for example, scientists and farmers use the APSIM model (McCown *et al.*, 1996) to analyse management options and help improve the farmers' and scientists' understanding of the soil–crop system (McCown *et al.*, 1998). In an application of APSIM to wheat farming systems on the Liverpool Plains, New South Wales, Turpin *et al.* (1998) found that yields were maximized when sufficient nitrogen was consistently applied to grow the 90 percentile yield with 13% protein.

Changed soil structure

Rundown in soil structure is a serious problem for dryland and irrigated cropping on Vertisols in Australia, and has sparked extensive research (e.g. Figs 16.6 and 16.7). Reduced aggregate stability of surface soil as a result of cropping can cause hydraulic conductivity of surface seals to fall to very low levels, as seen in Fig. 16.6. Maintenance of cover is essential on these soils to prevent the formation of seals with such low hydraulic conductivity (Freebairn *et al.*, 1986b). Recognition of the importance of cover and plant residues in reducing surface crusting is high among the farming community, prompting take-up of stubble retention in their farming systems. Ley rotations are also favourable for ameliorating surface crusting, because they add organic matter to the soil, create macro-porosity and help to re-aggregate the soil (Connolly *et al.*, 1998).

Compaction, as a result of wheel track pressure and smearing with tillage implements, is known to cause problems with infiltration and crop growth (McGarry, 1987; Gupta *et al.*, 1989). McGarry (1990) found a paddock with severe compaction precipitated a chain of events that led to a 73% reduction in cotton yield. Methods of ameliorating compaction (ley rotations to add organic matter to the soil, crops that encourage cracking) and reducing ongoing compaction (zero or minimum tillage, controlled traffic) are increasingly being adopted by farmers.

Rainfall-use efficiency and cropping management

Because of Australia's variable rainfall and high evaporative demand, there is widespread interest in making better use of whatever rain falls. This involves understanding more about the soil, particularly with respect to restrictions

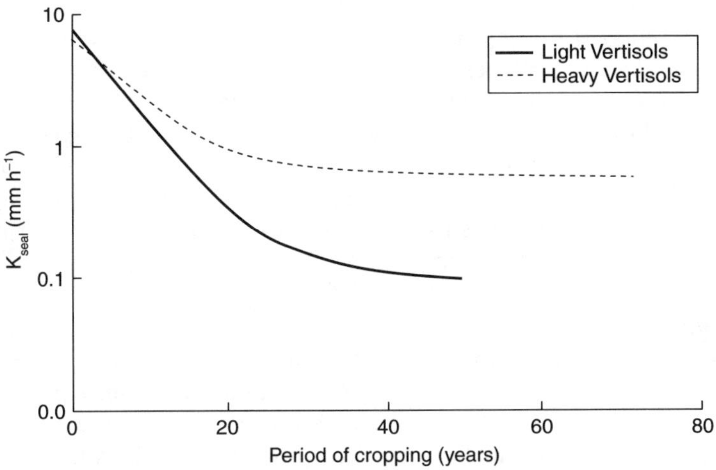

Fig. 16.6. Rundown in hydraulic conductivity of a surface crust of Vertisols in Queensland and New South Wales with period of cropping (Connolly *et al.*, 1997). Hydraulic conductivity was measured after 30 min of high-intensity simulated rainfall on bare soil. Hydraulic conductivity has declined because of rundown in organic matter and disruption of soil aggregates by tillage.

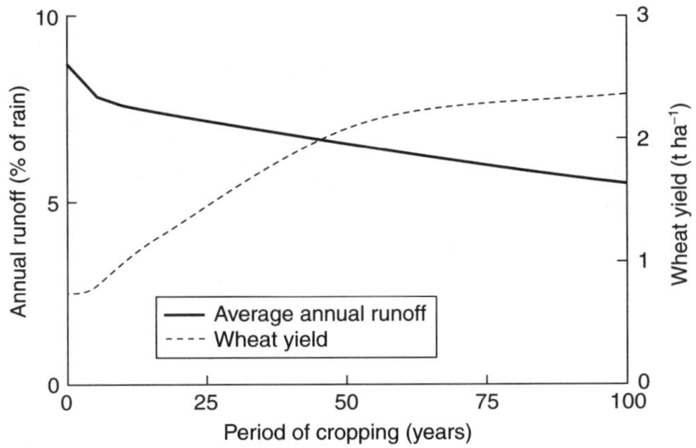

Fig. 16.7. Simulated effect of rundown in soil structure on average annual runoff and wheat yield for a wheat–summer fallow cropping system on the Heavy Vertisol in Fig. 16.6. Data are simulated with APSIM–SWIM. Runoff has increased because of increased surface crusting and reduced permeability of sub-surface layers. Reduced infiltration combined with changes to the water balance (particularly soil water storage over the fallow and evaporation) led to the decline in yield.

to crop roots and infiltration of rainfall, and designing cropping systems that account for local conditions. Typically, efficient cropping systems revolve around improved fallow and crop management.

Fallow management

Techniques to improve fallow management include increasing infiltration, hence reducing runoff, encouraging infiltrated water to move deeper into the soil profile where it is safe from evaporation, maintaining a weed-free fallow, minimizing disease and pest problems, and encouraging suitable seedbed conditions. A range of cropping systems have been developed that address these issues, but much remains to be learnt, particularly with respect to disease management and efficient operation of machinery. Research has often provided the impetus and support for improved fallow management systems, but practical management strategies are most often developed on farm by researchers and farmers working together (see Chapter 15).

Cropping options

Cropping options that are flexible, taking advantage of water stored in the soil as soon as it becomes available, show promise for improving efficiency of cropping (e.g. Wylie, 1999). Developing these options requires a sound knowledge of soil water-holding capacity, principles of soil water storage and monitoring, the climate and cropping options. Leading farmers in Australia are already actively involved in learning and monitoring programmes, and agribusiness is becoming more active. Tools such as Soil Matters (Dalgleish and Foale, 1998), HOWWET? (Freebairn *et al.*, 1996a) and the SoilPak system (e.g. Daniels *et al.*, 1994) are indicators of interest in this area.

Models have been used successfully for evaluating cropping options and developing improved crop rotations, at times directly in conjunction with farmers (e.g. simulation of effects of rainfall variability on wheat yields – FARMSCAPE, McCown *et al.*, 1998; Robinson and Freebairn, 1999). Even simple models have been used to good effect. Robinson *et al.* (1999) used a water-use efficiency model (a simple model relating rainfall to water uptake by crops) to analyse the feasibility of growing dryland cotton at St George, concluding that careful attention to accumulation of soil water over fallows and management of yield variability could allow profitable production of dryland cotton in a relatively dry area.

Environmental impact

Recently, much research has been directed at minimizing impacts of farming operations on off-farm areas, particularly rivers and water bodies. For example, recently the cotton industry has been involved in a joint study assessing the impact of pesticides on the riverine environment and developing best management practices for on-farm management of pesticides (Schofield *et al.*, 1998). However, cropping industries, in general, are under substantial pressure to reduce off-site movement of pesticides, nutrients and sediment, and research is still under way to improve the environmental performance of Australian cropping industries.

Grazing lands

Research in grazing lands has centred on development of systems that cater for uncertainty in rainfall, invasion by woody weeds and other pests and improved pasture productivity. Models have played and will continue to play a strong part in the development of improved management strategies (e.g. McKeon *et al.*, 1990). The 'Aussie Grass' project involves running a simulation of grass growth on a 5-km grid over the entire state of Queensland for the purposes of monitoring and evaluating rangeland condition and developing improved management strategies (DNR, 1999). The challenge for the future is in the development of more flexible strategies, based on model predictions, and the uptake of simulation results by farmers.

Using soil–crop models

Soil–crop models are useful for extending experimental data sets and are used in research on Vertisols. For example, once the properties of soils and crops are known and are represented in the model, the model can be used to see how successfully a new crop might grow in a new region. If the model suggests the crop will grow well, then some experimentation could be implemented to see how the crop performs in practice.

Soil–crop models are mathematical representations of the soil–crop system. Typically, these models represent the response of the crop to water and nutrient stress. Figure 16.8 shows what the APSIM (McCown *et al.*, 1996) model represents. APSIM considers all elements of the water balance (i.e. how rainfall is divided up into runoff, infiltration, transpiration, evaporation, drainage below the crop root zone and storage in the soil) as well as estimating erosion and tracking crop residue, nitrogen and carbon pools, crop growth and yield.

The models do not replace experimentation, because they do not necessarily simulate all of the processes that affect crop growth. Processes that most models do not simulate include disease and pests and other problems caused by compact soil, limiting nutrients other than nitrogen, etc. Experimental results are also needed to set up and parameterize the models and to test that the models are reliable.

Models help to integrate experimental research and extend results to different combinations of crops, soils or climates. The model can also be set up and run relatively quickly for a long climate record; experimentation takes

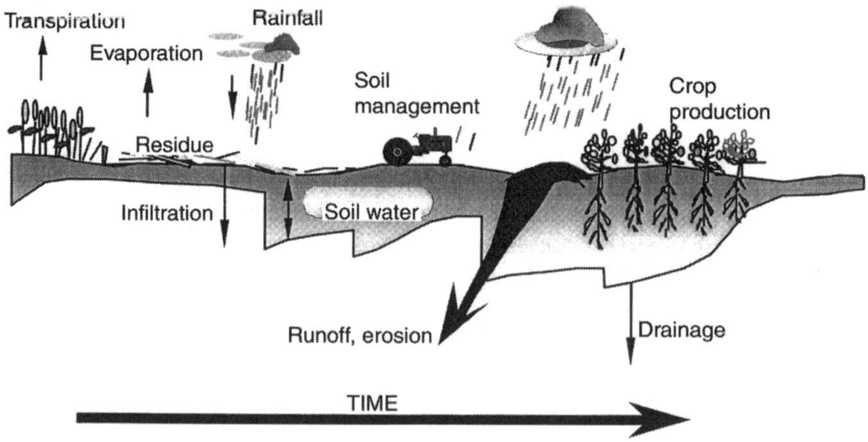

Fig. 16.8. A diagrammatic representation of the APSIM model.

time, field trials are expensive to operate and experimental records will never be as long as our weather records.

Using APSIM to Interpret and Add Value to a Chisumbanje Experimental Study

What we did

The APSIM model was used to simulate the effect on maize yields of water ponding in 'tied' furrows versus a flat surface ('on the flat'). Effects of these treatments on crop yield were measured in an experiment at the Chisumbanje Experimental Station, Zimbabwe (Jones *et al.*, 1989; Fig. 16.9). The experiment concluded that extra infiltration associated with water ponding in the tied furrows increased average maize yield by 45%, but response from year to year was highly variable. The observed differences in yields were not likely to be due to nutritional effects (Jones *et al.*, 1989).

We used the APSIM (McCown *et al.*, 1996) model to extend and compliment the results of the experiment. APSIM has been tested and applied to a range of management scenarios on Vertisol soils in Australia (e.g. Probert *et al.*, 1995; McCown *et al.*, 1996; Probert *et al.*, 1998; Robinson and Freebairn, 1999). A comparison of measured versus predicted yield for several crops in the Kingaroy area, Queensland, is shown in Fig. 16.10.

Unfortunately we did not have detailed rainfall records for the experimental period or a good description of soil properties. As a first estimate, we used general soil properties of Chisumbanje Vertisols described in Mandiringana and Hussein (1989) and used crop characteristics of a Zimbabwe maize variety. We did not have a long climate record for the Chisumbanje area, so we

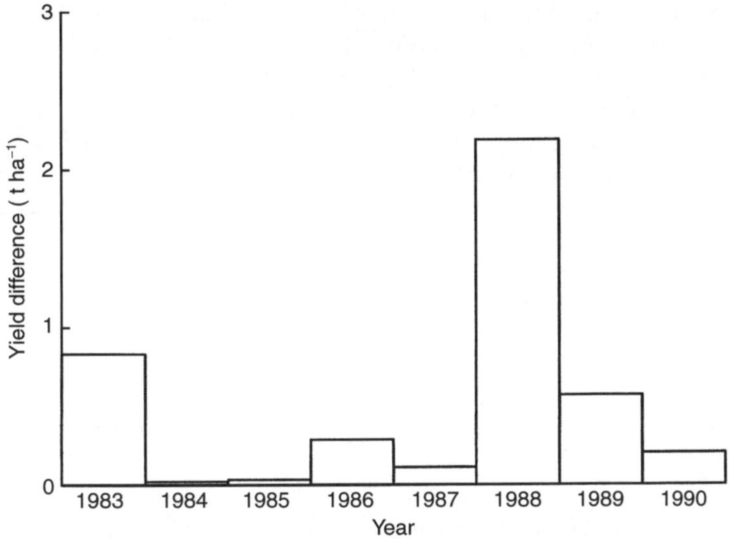

Fig. 16.9. Difference in grain yield between maize grown in 'tied' furrows and 'on the flat' in the experiment at the Chisumbanje Experimental Station, Zimbabwe (Jones *et al.*, 1989).

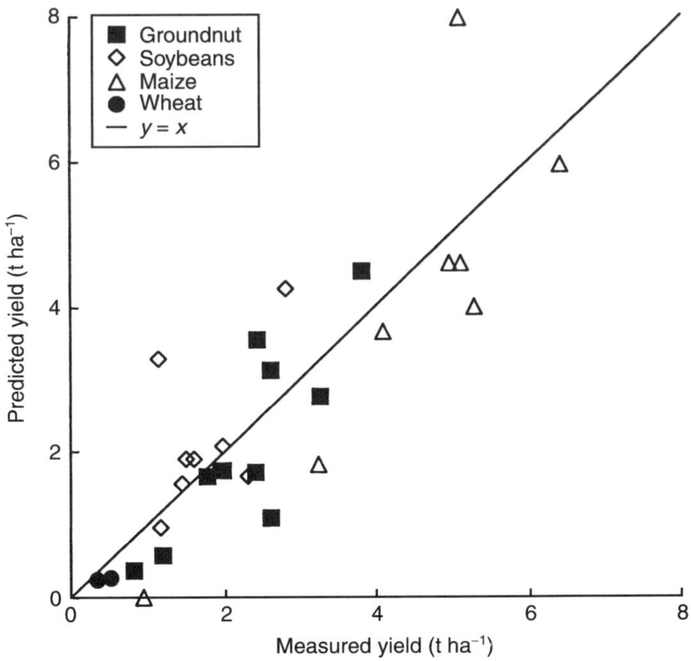

Fig. 16.10. Measured versus predicted yield for several crops near Kingaroy, Queensland, simulated with APSIM.

had to use a record for Bulawayo, several hundred kilometres to the west. We assumed furrows stored 10 mm of runoff.

Because we did not have detailed records for the experiment, a direct comparison of experimental yields with the model predictions was not meaningful. In this instance we use the model to illustrate how a model can be used to add value to the experimental results. The experiment operated from 1982 to 1990. This is a substantial experimental duration, but does not necessarily capture variability in the weather likely to be experienced over a longer time frame. The model simulations used a longer (40 year) weather record. We also simulated components of the water balance that were not measured, including runoff.

Results

Figure 16.11 shows the simulated yield of maize grown 'on the flat' for the 40-year weather record. Variability in simulated yield from year to year is large and the period covering the experimental work (the last 8 years) has a series of relatively low yields and failed crops. This raises a question as to how representative the experimental period was in the longer term. Figure 16.11 indicates that average yield over the experimental period was lower than the long-term average.

Variability in response to water ponding in furrows was also large (Fig. 16.12). Again, average response over the experimental period was different from over the longer term.

As well as simulating yield, APSIM is also effective in simulating changes to the various components of the water balance that were not studied in the experiment. Figure 16.13 shows average annual runoff for maize grown on the flat and in furrows. Furrows (storing up to 10 mm of runoff) reduced runoff to a third of the flat treatment. The changed runoff also affected the water

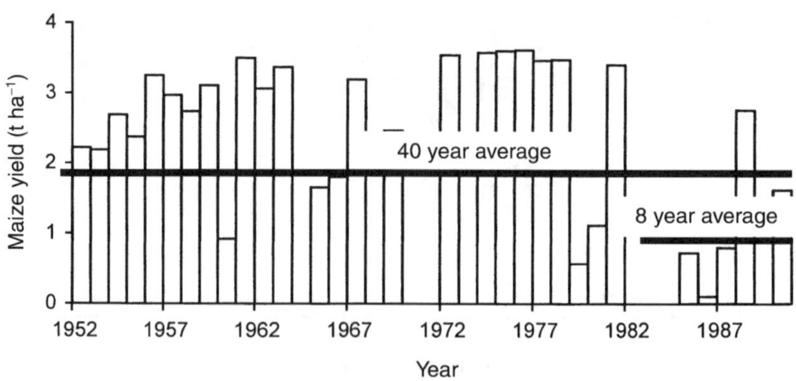

Fig. 16.11. Simulated grain yield of maize grown on the flat showing the large variability in yield from year to year.

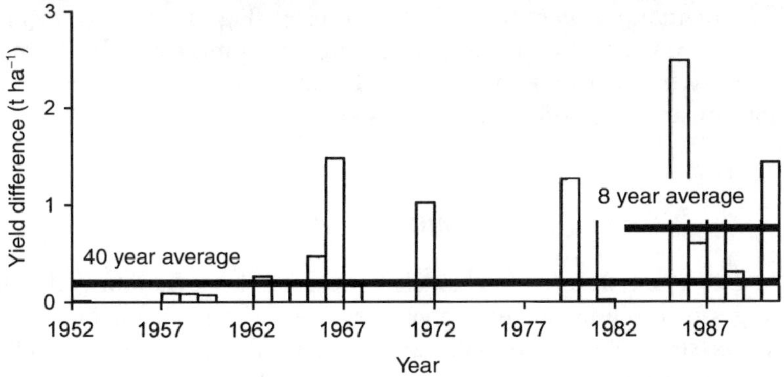

Fig. 16.12. Simulated difference in grain yield between maize grown on the flat and in furrows.

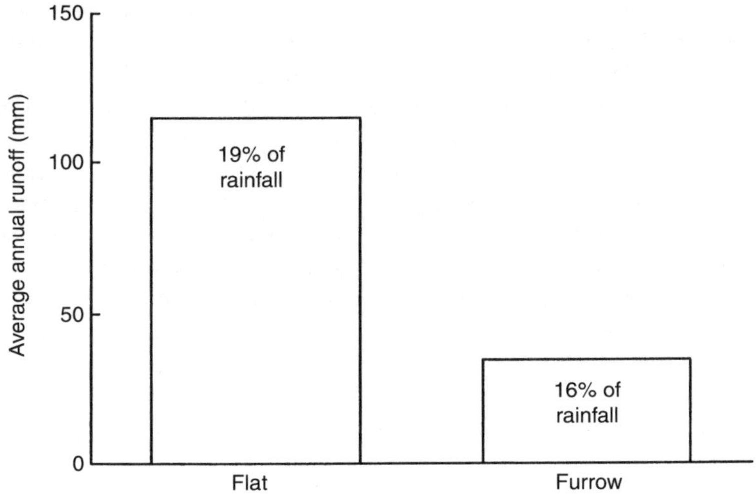

Fig. 16.13. Simulated difference in average annual runoff between maize grown on the flat and in furrows.

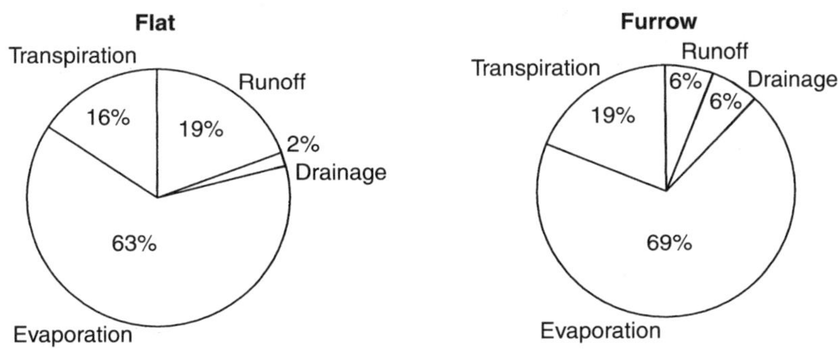

Fig. 16.14. Change in the complete water balance between maize grown on the flat and in furrows.

balance, including water available to the crop (Fig. 16.14). Transpiration increased by about 20%, causing an average yield increase of 0.2 t ha^{-1}, or 13%. However, because of the extra infiltration, losses to drainage increased threefold and evaporation increased by about 10%.

Discussion

This is one example of how models are useful for evaluating the effects of changes in a management/cropping system on other components of the soil–crop system. There are many other potential applications for soil–crop models, e.g. evaluating the risk–reward trade-offs for applying fertilizer or translating existing crops to new areas. In this example, the model did not replace the experimental data set. Rather, the data set was used to derive parameter values for the model. The model was then used to simulate a longer weather record and a greater range of variables (particularly the complete water balance). The simulation results can then be interpreted in the context of the experimental conclusions. The measured results can also be interpreted in relation to longer-term variability in the weather and crop growth.

It is important, though, to have confidence in the model output. This confidence can be built by ensuring the physical system is well represented in the model and by testing the model's ability to reproduce measured yield, soil water, runoff, etc. In areas where long-term data are not available, testing and application of the model are more difficult. In the example given here, we applied the model to a scenario in Zimbabwe using a very limited database. We could do this because the model has previously been well tested for similar soils and conditions in Australia and elsewhere. Once more local data are available, though, possibilities for exploring management options are greatly increased.

Summary

Australia has a large area of highly productive Vertisols used for cropping and grazing. But a number of land-use-related problems pose challenges when farming these soils. After up to 70 years of cropping, overgrazing, erosion and rundown in soil physical properties have placed pressure on the soil resource. Highly variable rainfall means water stress is a common limitation to crop and pasture production. And, increasingly, impacts of farming activities on the environment are coming under closer inspection.

Land managers have responded to these pressures by implementing soil conservation works and management systems that minimize tillage, maximize surface cover, minimize pesticide and nutrient export off-farm, maintain or improve organic matter levels and make the most efficient use of rainfall. But there is still much to be learnt about how best to apply these strategies in an economically viable fashion. The challenge for land managers is to make best use of knowledge banks and tools, including computer models, to develop flexible, specific solutions to land management problems on Vertisols.

References

Ahmad, N. (1996) Occurrence and distribution of Vertisols. In: Ahmad, N. and Mermut, A. (eds) *Vertisols and Technologies for Their Management.* Developments in Soil Science No. 24. Elsevier, Amsterdam, pp. 1–41.

Campbell, A. (1992) National landcare facilitator. In: *Annual Report.* Department of Industry and Energy, Canberra.

Connolly, R.D., Freebairn, D.M. and Bridge, B.J. (1997) Change in infiltration characteristics associated with cultivation history of soils in south-eastern Queensland. *Australian Journal of Soil Research* 35, 1341–1358.

Connolly, R.D., Freebairn, D.M. and Bell, M.J. (1998) Change in soil infiltration with leys in south-eastern Queensland. *Australian Journal of Soil Research* 36, 1057–1072.

Dalal, R.C. and Mayer, R.J. (1986a) Long-term trends in fertility of soils under continuous cultivation and cereal cropping in Southern Queensland. I Overall changes in soil properties and trends in winter cereal yields. *Australian Journal of Soil Research* 24, 265–279.

Dalal, R.C. and Mayer, R.J. (1986b) Long-term trends in fertility of soils under continuous cultivation and cereal cropping in Southern Queensland. II Total organic carbon and its rate of loss from the soil profile. *Australian Journal of Soil Research* 24, 281–292.

Dalal, R.C., Strong, W.M., Weston, E.J. and Gaffney, J. (1991) Sustaining multiple production systems. 2. Soil fertility decline and restoration of cropping lands in sub-tropical Queensland. *Tropical Grasslands* 25, 173–180.

Dalal, R.C., Strong, W.M., Weston, E.J., Cooper, J.E., Lehane, K.J., King, A.J. and Chicken, C.J. (1995) Sustaining productivity of a Vertisol at Warra, Queensland, with fertilizers, no-tillage, or legumes. 1. Organic matter status. *Australian Journal of Experimental Agriculture* 35, 903–913.

Dalal, R.C., Strong, W.M., Weston, E.J., Cooper, J.E., Wildermuth, G.B., Lehane, K.J., King, A.J. and Holmes, C.J. (1998) Sustaining productivity of a Vertisol at Warra, Queensland, with fertilizers, no-tillage, or legumes. 5. Wheat yields, nitrogen benefits and water-use efficiency of chickpea–wheat rotation. *Australian Journal of Experimental Agriculture* 38, 489–501.

Dalgleish, N. and Foale, M. (eds) (1998) *Soil Matters. Monitoring Soil Water and Nutrients in Dryland Farming.* Agricultural Production Systems Research Unit, Toowoomba, Australia.

Daniels, I., Brown, R. and Deegan, L. (eds) (1994) *Northern Wheatbelt Soilpak. A Soil Management Package for Dryland Cropping in the Summer Rainfall Zone.* NSW Agriculture, Tamworth, Australia.

DNR (Department of Natural Resources) (1999) Web site: http://www.dnr.qld.gov.au/longpdk/agrass/index.html

Freebairn, D.M., Ward, L.D., Clarke, A.L. and Smith, G.D. (1986a) Research and development of reduced tillage systems for Vertisols in Queensland, Australia. *Soil and Tillage Research* 8, 211–229.

Freebairn, D.M., Wockner, G.H. and Silburn, D.M. (1986b) Effects of catchment management on runoff, water quality and yield potential from a Vertisol. *Agricultural Water Management* 12, 1–19.

Freebairn, D.M., Littleboy, M., Smith, G.D. and Coughlan, K.J. (1991) Optimising soil surface management in response to climate risk. In: Muchow, R.C. and Bellamy, J.A. (eds) *Proceedings International Symposium on Climate Risk in Crop Production: Models and Management for the Semiarid Tropics and Subtropics.* CAB International, Wallingford, UK, pp. 283–305.

Freebairn, D.M., Hamilton, A.N., Glanville, S.F. and Dimes, J.P. (1996a) *HOWWET? – a Software Package for Predicting Fallow Water Storage and N Mineralisation.* Queensland Department of Primary Industries, Toowoomba, Queensland.

Freebairn, D.M., Loch, R.J. and Silburn, D.M. (1996b) Soil erosion and soil conservation for Vertisols. In: Ahmad, N. and Mermut, A. (eds) *Vertisols and Technologies for their Management.* Elsevier, New York, pp. 303–362.

Gupta, S.C., Sharma, P.P. and Defranchi, S.A. (1989) Compaction effects on soil structure. *Advances in Agronomy* 42, 311–338.

Holford, I.C.R. (1980) Effects of duration of grazed lucerne on long-term yields and nitrogen uptake of subsequent wheat. *Australian Journal of Agricultural Research* 31, 239–250.

Holford, I.C.R. and Crocker, G.J. (1997) A comparison of chickpeas and pasture legumes for sustaining yields and nitrogen status of subsequent wheat. *Australian Journal of Agricultural Research* 48, 305–315.

Jones, E., Nyamudeza, P. and Busangavanye, T. (1989) Rainfed cropping and water conservation on Vertisols in the southeastern lowveld of Zimbabwe. In: Ahn, P.M. and Elliot, C.R. (eds) *Vertisol Management in Africa.* IBSRAM Proceedings No. 9. IBSRAM, Bangkok, Thailand, pp. 133–142.

Knisel, W.G. (ed.) (1980) *CREAMS: a Field Scale Model for Chemicals, Runoff, and Erosion from Agricultural Management Systems.* US Department of Agriculture, Conservation Research Report No. 26. USDA, Washington, DC.

Lane, L.J. and Nearing, M.A. (eds) (1989) *USDA – Water Erosion Prediction Project: Hillslope Profile Model Documentation.* NSERL Report No. 2. USDA-ARS National Soil Erosion Research Laboratory, West Lafayette, Indiana.

Littleboy, M., Silburn, D.M., Freebairn, D.M., Woodruff, D.R. and Hammer, G.L. (1989) *PERFECT. A Computer Simulation Model of Productivity Erosion Runoff Functions to Evaluate Conservation Techniques.* Queensland Department of Primary Industries, Brisbane.

Mandiringana, O.T. and Hussein, J. (1989) Properties and management of the Chisumbanje Vertisols. In: Ahn, P.M. and Elliot, C.R. (eds) *Vertisol Management in Africa.* IBSRAM Proceedings No. 9. IBSRAM, Bangkok, Thailand, pp. 119–132.

McCown, R.L., Hammer, G.L., Hargreaves, J.N.G., Holzworth, D.P. and Freebairn, D.F. (1996) APSIM: a novel software system for model development, model testing and simulation in agricultural systems research. *Agricultural Systems* 50, 255–271.

McCown. R.L., Carberry, P.S., Foale, M.A., Hochman, Z., Coutts, J.A. and Dalgleish, N.P. (1998) The FARMSCAPE approach to farming systems research. In: Michalk, D.L. and Prately, J.E. (eds) *Agronomy – Growing a Greener Future. Proceedings of the 9th Australian Agronomy Conference.* Charles Sturt University, Wagga Wagga, pp. 633–666.

McGarry, D. (1987) The effects of soil water content during land preparation on soil physical condition and cotton growth. *Soil and Tillage Research* 9, 287–302.

McGarry, D. (1990) Soil compaction and cotton growth on a Vertisol. *Australian Journal of Soil Research* 28, 869–877.

McKeon, G.M., Day, K.A., Howden, J.J., Mott, D.M., Scattini, W.J. and Weston, E.J. (1990) Northern Australian savannas: management for pastoral production. *Journal of Biogeography* 17, 355–372.

Ockerby, S.E., Lyons, D.J., Keefer, G.D., Blamey, F.P.C. and Yule, D.F. (1993) Irrigation frequency and nitrogen fertilizers modify cotton yield at Emerald, Central Queensland. *Australian Journal of Agricultural Research* 44, 1389–1402.

Probert, M.E., Keating, B.A., Thompson, J.P. and Parton, W.J. (1995) Modelling water, nitrogen, and crop yield for a long-term fallow management experiment. *Australian Journal of Experimental Agriculture* 348, 941–950.

Probert, M.E., Carberry, P.S., McCown, R.L. and Turpin, J.E. (1998) Simulation of legume–cereal systems using APSIM. *Australian Journal of Agricultural Research* 49, 317–327.

QDPI (1966) *Soil Conservation Handbook.* Queensland Department of Primary Industries, Soil Conservation Branch.

Robinson, B. and Freebairn, D. (1999) Long-term changes in rainfall and potential wheat yields at St George. In: *Confarm 21. The Second National Conservation Farming and Minimum Tillage Conference.* Conservation Farmers Inc. Qld, Moree Conservation Farmers Association NSW, Toowoomba, Australia, pp. 235–240.

Robinson, B., Freebairn, D., Castor, M. and Glanville, S. (1999) Effects of rainfall availability and variability on dryland cotton production. In: *Confarm 21. The Second National Conservation Farming and Minimum Tillage Conference.* Conservation Farmers Inc. Qld, Moree Conservation Farmers Association NSW, Toowoomba, Australia, pp. 221–229.

Rummery, G. and Coleman, B. (1999) Does conservation farming pay? In: *Confarm 21. The Second National Conservation Farming and Minimum Tillage Conference.* Conservation Farmers Inc. Qld and Moree Conservation Farmers Association NSW, Toowoomba, Australia, pp. 50–54.

Schofield, N., Edge, V. and Moran, R. (1998) Minimising the impact of pesticides on the riverine environment using the cotton industry as a model. *Water* Jan/Feb, 37–40.

Strong, W.M. and Cooper, J.E. (1995) Application of anhydrous ammonia or urea during the fallow period for winter cereals on the Darling Downs, Queensland. I. Effect of time of application on soil mineral N at sowing. *Australian Journal of Soil Research* 30, 695–709.

Thorburn, P.J., Coughlan, K.J., Gardner, E.A. and Yule, E.A. (1989) Soil water restrictions to land use of Vertisols in Queensland, Australia. In: *Vertisol Management in Africa. Proceedings of a Workshop Held at Harare, Zimbabwe.* IBSRAM Proceedings No. 9. IBSRAM, Bangkok, Thailand, pp. 77–96.

Turpin, J.E., Hayman, P.T., Marcellos, H. and Freebairn, D.F. (1998) Nitrogen fertiliser decisions for wheat on the Liverpool Plains NSW 1. Should farmers consider paddock history and soil tests? In: Michalk, D.L. and Pratley, J.E. (eds) *Proceedings of the 9th Australian Agronomy Conference,* Wagga Wagga, July 1998. Charles Sturt University, Wagga Wagga, pp. 761–764.

Viessman, W., Jr, Knapp, J.W., Lewis, G.L. and Harbaugh, T.E. (1977) *Introduction to Hydrology.* Harper and Row, New York.

Williams, J.R., Dyke, P.T., Fuchs, W.W., Bensen, V.W., Rive, O.W. and Taylor, E.D. (1990) *EPIC – Erosion/Productivity Impact Calculator. 2. User Manual.* US Department of Agriculture Technical Bulletin No. 1768. USDA, Washington, DC.

Wischmeier, W.H. and Smith, D.D. (1978) *Predicting Rainfall Erosion Losses. A Guide to Conservation Planning.* Agriculture Handbook No. 537. US Department of Agriculture, Washington, DC.

Wylie, P. (1999) Planning profitable rotations in northern Australia. In: *Confarm 21. The Second National Conservation Farming and Minimum Tillage Conference.* Conservation Farmers Inc. Qld, Moree Conservation Farmers Association NSW, Toowoomba, Australia, pp. 146–150.

The Vertisols of Texas 17

K.N. POTTER[1] AND T.J. GERIK[2]

[1]USDA-ARS, Grassland Soil and Water Research Laboratory, Temple, Texas, USA; [2]Texas A&M Experiment Station, Blackland Research Center, Temple, Texas, USA

Key words: USA, Texas, Vertisols, yields, economic analysis, infiltration, rainfall simulator

Introduction

Vertisols occur on about 6.5 million ha in Texas – over half of the Vertisols known within the US. Vertisols occur in all portions of the state and are the dominant soils in both the central Texas Blackland Prairies and Coastal Prairie Major Land Resource Areas (Newman, 1986). The major areas of Vertisols extend from the northern border southward in a narrow belt and along the Texas Gulf Coast region (Fig. 17.1). These areas include both densely populated urban and sparsely populated rural areas. Over half of the population of Texas lives either on Vertisols or vertic soils (*Texas Almanac*, 1998). Cities in Texas which occur on Vertisols include Austin, Beaumont, Corpus Christi, Dallas, Fort Worth, Houston, San Antonio and Waco as well as many smaller cities and towns. A large part of the highway system has also been constructed on Vertisols, including Interstate Highway 35, a major route extending from Mexico to Canada.

Urbanization resulting from the major cities and industries building along Interstate Highway 35 increases every year. From 1972 to 1992, the average annual rate of urbanization in the Blackland Prairie was 0.8% (Allen and Maier, 1993). Each year about 52,000 ha of land changed from largely agricultural land use to housing, roads and industrial uses.

Rural land uses have changed substantially in the last 60 years, with a decrease in the amount of land harvested for crops being the most noticeable change. From the 1930s to the present, over 80% of the area in rural Blackland Prairie counties was used for some type of farming activity. However, the percentage of farmland harvested for crops has declined sharply from about 75% in 1925 to about 20% in 1987 (US Department of Agriculture,

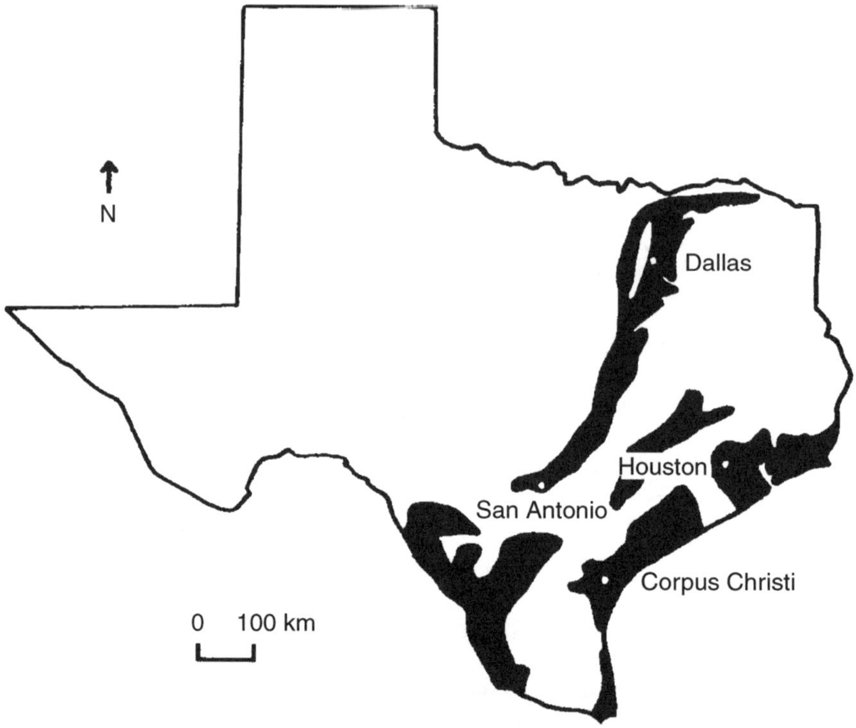

Fig. 17.1. Distribution of Vertisols in Texas, USA (from Coulombe *et al.*, 1996).

1993). The percentage of farmland harvested was similar in 1997, but 24–52% of the harvested land was in hay, silage or other forms of livestock fodder (US Department of Agriculture, 1998). Most of the land that has been taken out of cropland has been put into pasture. Livestock production has therefore become a major land use on Vertisols in the Blackland Prairie region. Grazing is produced in several different ways, including native and improved pastures, and grazing of small grains in place of grain harvest. Similar land-use trends have occurred on other Vertisols in Texas.

Approximately 1.04 million ha of Vertisols are cropped for harvest in Texas. Major crops grown in the Blackland Prairie include grain sorghum (*Sorghum bicolor*), wheat (*Triticum aestivum*), maize (*Zea mays*) and cotton (*Gossypium hirsutum* L.) (US Department of Agriculture, 1998). For the Coastal Bend region, major crops include maize, wheat, cotton, rice (*Oryza sativa*) and soybeans (*Glycine max*) (US Department of Agriculture, 1998).

Texas Vertisols receive an average annual precipitation ranging from 700 to 1000 mm. Most of the rain occurs in the autumn and spring of the year with annual summer droughts. The most common soil temperature regime is thermic.

Sustainable Management Research for Vertisols

The remainder of this chapter will discuss the results of soil management research on Vertisols conducted by the US Department of Agriculture, Agricultural Research Service, Grassland, Soil and Water Research Laboratory and the Texas A&M Agricultural Experiment Station located on the Blackland Prairie near Temple, Texas. The predominant soil at Temple is a Houston Black clay (fine montmorillonitic, thermic Udic Haplusterts). Temple is located at 31° north latitude and 97° west longitude with a mean elevation of about 213 m. The Houston Black clay soil contains 4% sand, 39% silt and 56% clay in the surface 30 cm. Mean annual rainfall is 863 mm. Most rain occurs between October and July, with the driest period occurring from July to September.

Common tillage practices used by farmers in Texas on Vertisols are both intensive and extensive. The predominant 'conventional tillage' during the 1990s on Blackland Prairie Vertisols has been a disc and chisel tillage system. This system typically would include: stalk shredding, tandem disc cultivation to cut and partially bury crop residues, primary tillage with a chisel plough, further tillage with a tandem disc harrow, and seedbed preparation with a field cultivator soon before planting. The crop may or may not be cultivated, but in nearly all cases growing-season weed control is by herbicides. Inorganic fertilizers are the general rule. Problems often cited when cropping Vertisols include low infiltration rates when soils are wet, resulting in susceptibility to large amounts of runoff and erosion, adhesive soils when wet and hard soils when dry, which restrict the potential opportunities for tillage, fertilization and planting. Also, Vertisols are susceptible to compaction when field operations are conducted at high soil water content.

Research Findings on Central Texas Vertisols

Morrison *et al.* (1990) described a conservation farming system that was designed to address perceived problems encountered when cropping Vertisols. The proposed farming system was similar to the broadbed and furrow system being used in Ethiopia (Asamenew *et al.*, 1988). Raised beds, 0.15 m high and 1.5 m wide, provided areas for three to eight crop rows. The beds were separated by 0.5 m wide furrows that serve both as restricted traffic lanes and as surface water drainage channels to prevent surface ponding of water during periods of excess rainfall. For experimental purposes, the beds were maintained in two ways. The first involved the beds being tilled out and rebuilt annually. The second treatment involved maintaining the beds from year to year with no-tillage management practices, which maintained a crop-residue cover. With the no-tillage management, the furrows were tilled every 3 years to maintain the shape of the bed. Weed control was maintained with herbicides so that the soil surface was not disturbed by tillage. Long-term experiments were established using this system to evaluate the effect of management system on crop growth and productivity and the effect of no-tillage management on Vertisol soil properties. A winter wheat–grain sorghum–maize crop rotation was used to determine the effects of annual tillage and no-tillage management systems. Each crop was present each year of the study.

Crop yields

Crop yields of no tillage and annually tilled beds were similar for the initial 6 years of the study, or two cycles of a wheat–maize–grain sorghum crop rotation (Fig. 17.2). Crop yields increased when low rates of fertilizer were applied. However, wheat yields did not increase with additional increase in rates of fertilizer. Maize yields increased with increasing rates of fertilization, to rates as high as 168 kg N ha^{-1} and 37 kg P ha^{-1}.

In a separate but similar study, five levels of tillage intensity were tested for 3 years to determine the effects on growth and yield of maize and grain sorghum on a Houston Black clay soil (Potter *et al.*, 1996). Tillage intensity treatments included: chisel plough with secondary tillage; disc only; no tillage with residue rakes at planting; no tillage with mid-season cultivation; and no

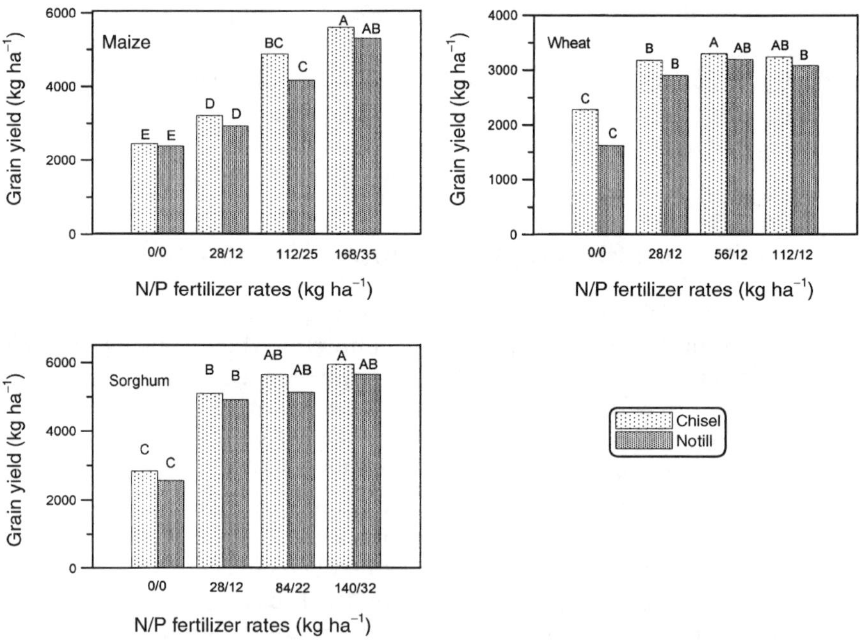

Fig. 17.2. Grain yields as a function of fertilizer rates. Comparisons within a treatment across fertilizer rates with the same letter are not significantly different at $P = 0.05$ by Duncan's multiple range test.

tillage. The no-tillage treatments were established on sites that had been in continuous no tillage for 10 years, allowing for no-tillage management effects to be expressed. Maize plant populations were greater in tilled treatments than in no-tillage treatments in 2 of the 3 years. Above-ground biomass production of maize was lower in no-tillage treatments early in the growing season, but by anthesis differences among treatments were not significant. Maize yielded 940 kg ha^{-1} more when the soil was tilled than with no tillage. Plant populations accounted for much of the difference in maize grain yields between tillage systems, with low plant populations of no tillage restricting yield in some years. Grain sorghum populations were not affected consistently by tillage intensity, and biomass production was less sensitive to tillage intensity than maize. Grain sorghum yields were generally as large or larger in no-tillage treatments as compared with the tilled treatments.

Economic analysis

An economic analysis, using labour, machinery prices and herbicide costs representative of the 1990s, indicated that no tillage was more cost-efficient than chisel tillage (Harman *et al.*, 1996). Total variable cost (i.e. cost of seed, fertilizer, herbicides, fuel, labour and repairs) was reduced by US$40 ha^{-1} $year^{-1}$ for no tillage as compared with chisel tillage. Machinery fixed costs were also reduced by US$22 ha^{-1} $year^{-1}$ for no tillage compared with chisel tillage with the broadbed and furrow system and a wheat–maize–grain sorghum rotation. The largest difference since the 1980s is in the lower cost of herbicides. The reduced cost of production compensated for the reduced yields associated with no tillage after 10 years of a continuous no-tillage management system.

Management Effects on Soil Properties

To be sustainable, management practices must not result in detrimental soil conditions and soil quality must be maintained. Soil compaction and reduced water infiltration are common in degraded Vertisols (Cook *et al.*, 1992; Seiny-Boukar *et al.*, 1992) and may limit crop production.

Water infiltration

Rainfall simulator studies were conducted on a Houston Black clay soil that had been under continuous management for 4 years using the broadbed and furrow system (Potter *et al.*, 1995). Simulated rainfall was imposed on four tillage and residue combinations. Tillage regimes included no tillage except for reshaping the bed shoulders every 3 years, and annual chisel tillage with complete bed reformation. Crop residues from the previous winter wheat crop were removed from some no-tillage plots to provide a no-tillage/low-residue treatment. The residue was placed on some chisel tillage plots to provide a tilled/high-residue treatment. The percentage residue cover was over 90% in the no-tillage/high-residue and chisel-tillage/high-residue treatments. In the no-tillage/low-residue (28% cover) and chisel tillage/low-residue (39% cover) plots, percentage residue cover was higher than expected. However, the residue on the low-residue plots consisted of small particles, mostly chaff, which adhered to the soil surface. Two initial soil water contents were studied, a dry condition (0.12 $m^3\ m^{-3}$ volumetric soil water content) and a wet condition (0.48 $m^3\ m^{-3}$ volumetric soil water content), on the same areas. Water infiltration data during the dry run had a general pattern in all tillage/residue combinations of very large initial infiltration rate, followed by a rapid decline in infiltration rate. In most cases, a lower, nearly constant rate was established eventually. Similar results were reported for Vertisols in Australia with small (1 m^2) plots, and large (18 m^2) plots with rainfall simulators, and 1-ha watersheds with natural rainfall (Freebairn *et al.*, 1984). A large portion of the water infiltration occurred before any runoff was measured (Table 17.1). Several authors have noted the importance of this 'initial infiltration' in recharging the soil profile with water in Vertisols (Freebairn *et al.*, 1984; Bouma and Loveday, 1988; Coughlan *et al.*, 1989). In the dry condition, initial infiltration accounted for 100% of the cumulative infiltration during the initial 30 min rain in both chisel tillage and no tillage with large amounts of surface residue. With low amounts of residue, initial infiltration accounted for 65 and 89% of cumulative infiltration during the initial 30 min in both the chisel tillage and no-tillage sites, respectively.

During the wet rainfall simulation, residue amount significantly affected the amount of initial infiltration (Table 17.1). With low amounts of residue, initial infiltration amounts were very small, accounting for 5% of cumulative infiltration during the initial 30 min in both tillage treatments. With large amounts of surface residue, initial infiltration accounted for 53 and 29% of cumulative infiltration during the initial 30 min for tillage and no-tillage plots, respectively.

Surface residue cover also had a major effect on final water infiltration rates in both the dry and the wet runs. As expected, final infiltration rates were greater during the dry runs than in the wet runs (Table 17.1). The chisel tillage sites with residue never obtained a constant rate during the time of the dry run. In contrast, the chisel tillage plots with low residue had the lowest final rates of all treatments. Large amounts of surface residue cover resulted in significantly higher final infiltration rates during the wet runs in both chisel tillage and no-tillage plots (Table 17.1). Final infiltration rates averaged 56 mm h^{-1} for the chisel tillage and 45 mm h^{-1} for the no-tillage plots with high rates of surface mulch compared with 20 and 28 mm h^{-1} for the respective low-residue plots. Soil disturbance from tillage was not a significant factor in determining final rates during the wet run rainfall simulations.

Ponded and tension infiltrometers were used to determine soil hydraulic properties after 8 years of annual tillage and no tillage in broadbeds and furrows (Potter *et al.*, 1995). Tension infiltrometers measure infiltration rates at water pressures that are negative with respect to atmospheric pressure. Ponded infiltrometers by definition require a positive water pressure. Mean water infiltration rates for crop beds and furrows are illustrated in Fig. 17.3.

Table 17.1. Summary of rainfall simulator infiltration data.

	Tilled/ low residue	Tilled/ high residue	No tillage/low residue	No tillage/high residue
Initial infiltration (mm)				
Dry run	32.4 (13)	129 (41)	54 (26)	75 (45)
Wet run	1.1 (0.9)	31 (21)	1.4 (0.9)	12 (2)
Final infiltration rate (mm h^{-1})				
Dry run	28 (9)	–	58 (20)	78 (3)
Wet run	20 (8)	56 (3)	28 (8)	45 (24)
30-min cumulative infiltration (mm)				
Dry run	50 (6)	62 (9)	61 (5)	66 (7)
Wet run	21 (10)	58 (9)	28 (5)	42 (6)

Values are mean (n = 4) and standard deviation.

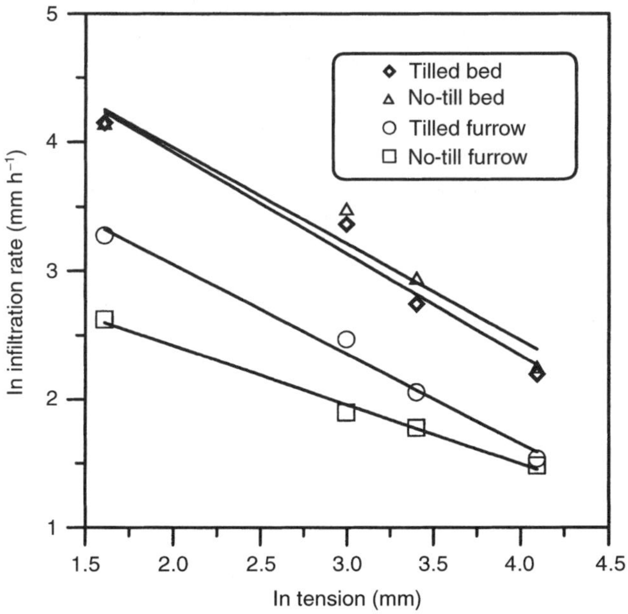

Fig. 17.3. Natural log of the unconfined infiltration rate vs. the natural log of tension for trafficked furrows and untrafficked beds (Potter *et al.*, 1995).

Infiltration rates were lower in the traffic furrows than in the crop beds. Although reduced to a greater extent at the no-tillage site, which had been trafficked for 8 years without deep tillage, infiltration rates within furrows were not statistically different between management systems probably because of the traffic that occurred during the cropping season. Low infiltration rates in the furrow help drain excess surface water rapidly from the field. Surface drainage is especially desirable in the spring when the soil is usually saturated. Water infiltration rates were not significantly different between tillage and no-tillage crop beds. This implies that similar pore size distributions and pore continuity were present with both management systems. Infiltration measurements were made about 1 year after tillage and after a wheat crop had been grown on this site. The soil had been dried by evapotranspiration by the previous wheat crop. Self-mulching and deep cracking of the soil had occurred during the growing season. However, at the time of the tension infiltrometer measurements, the soil had been rewetted so that cracking was not apparent at the soil surface. The crop canopy and surface residue also prevented the development of a substantial surface seal.

Sediment losses

In its native condition, there was little erosion in the Blacklands Region due to the protection given by native grasses. However, when the native grasses were broken out and the land was used to produce crops, erosion became a serious problem. In the 1920s over 70% of the Blackland Vertisols were tilled to produce crops, using the inversion tillage practices common at that time. A common rotation in the early 1940s, 2 years of cotton and 3 years of small grain, averaged sediment losses of 35 t ha^{-1} compared with 0.5 t ha^{-1} from a native grass meadow of similar slope (Richardson, 1993). Later it was determined that annual erosion losses were reduced by 75% using structural devices and practices such as terraces, contour cultivation and grassed waterways.

Research in recent years has demonstrated that no-tillage or residue-conserving minimum tillage practices successfully controlled erosion on Vertisols (Potter *et al.*, 1995). Surface residue cover was highly effective in reducing sediment losses from the raised beds. Mean sediment losses were nine to 15 times greater from the low-residue plots than from the high-residue plots (Table 17.2). Similar reductions in sediment loss with no tillage compared with annual tillage have been reported in watershed studies (Chichester and Richardson, 1992). While not statistically different, the no-tillage low-residue plots tended to have lower sediment losses than the tilled low-residue plots in the dry runs. Losses were similar between the low-residue treatments in the wet runs. Most of the sediment losses appeared to originate from the shoulder of the bed, with sediment slaking into the furrow. The high-residue plots did not appear to slake into the furrow. From visual observations, it appeared that weathered beds in tilled plots exhibited similar characteristics, with 1-year-old beds often eroded by natural rainfall to the first row of wheat stubble in the bed. Sediment that eroded into the furrows between beds was lost from the field because of rapid surface drainage from the fields.

Bulk density

Mean soil bulk density values for conventional tillage and no tillage were similar near the surface (0–4 cm) for the Houston Black clay (Fig. 17.4). At depths of 4–15 cm, the no-tillage soil had greater bulk density values than with chisel tillage. Below 15 cm, soil bulk density values were similar.

Soil organic carbon

Soil organic carbon (SOC) content has long been recognized as one indicator of soil quality. Intensive agricultural practices and resulting erosion and organic carbon oxidation have resulted in large losses in SOC in agricultural soils as compared with native prairie soils. The organic carbon concentration in the surface 0.6 m of a tilled Vertisol was greatly reduced compared with that of a

Table 17.2. Total sediment losses with 30 min runoff.

	Total sediment losses ($g\ m^{-2}$)			
	Tilled/low residue	Tilled/high residue	No tillage/low residue	No tillage/high residue
Dry	461 (205)	30 (27)	264 (134)	25 (24)
Wet	422 (195)	34 (24)	411 (48)	45 (30)

Values are mean (n = 4) and standard deviation.

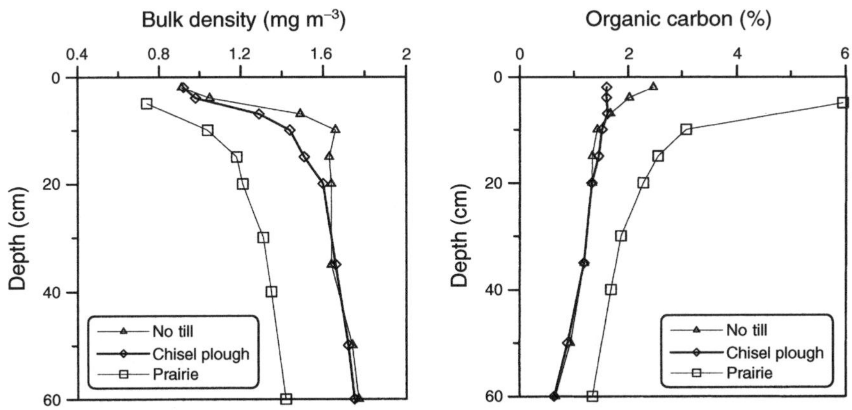

Fig. 17.4. Soil bulk density and soil organic carbon concentration distribution in tilled, no-tillage and native prairie Vertisols.

native prairie (Fig. 17.4). Carbon accumulation rates in soils under no tillage or conservation tillage reported in the literature have varied widely, ranging from below zero to 1.3 t C ha^{-1} $year^{-1}$ (Reicosky *et al.*, 1995). In the USA, the greatest increases in SOC have been reported in the colder, northern regions, and even there differences in SOC content due to management take several years to become measurable.

In recent experiments on Texas Vertisols, SOC concentration differences between annually chisel-tilled soils and no-tillage management after 10 years of continuous management were reported in the surface 20 cm (Fig. 17.4) (Potter *et al.*, 1998). A significantly greater mass of SOC was present in the no-tillage soils than in the tilled soils (Table 17.3). Fertilizer rate had no effect on total SOC. The rate of carbon sequestration in the surface 20 cm of soil resulting from conversion from chisel tillage to no tillage was 300 kg C ha^{-1} $year^{-1}$ after maize (Potter *et al.*, 1998). In comparison, net SOC increased at a rate of 447 kg C ha^{-1} $year^{-1}$ after returning degraded cropland to grass production for periods ranging from 6 to 60 years (Potter *et al.*, 1999).

Future Plans

Controlled-traffic and residue management cropping systems on Vertisols have not been well accepted in Texas to this date. The lack of acceptance by growers is probably due to insufficient education regarding the economic and environmental benefits of conservation tillage practices. Research efforts are continuing and expanding to include topics such as ultra-narrow row crop production systems and row-zone tillage.

Crops grown in ultra-narrow row configuration (e.g. with rows spaced less than 0.5 m) form a complete canopy quickly and more effectively intercept radiant energy. Evaporation losses of dryland sorghum in central Texas were reduced by 50% and yields increased by 20% as row spacing decreased from 1 m to 0.5 m (Adams *et al.*, 1976; Arkin *et al.*, 1978). In the 1970s, farmers did not have the planting and harvesting equipment or weed control technologies to apply this technology to their farms. These problems no longer exist today. Excellent weed control technologies using herbicide-resistant transgenic varieties and exciting new chemistries are rapidly replacing the plough. Advanced planting and harvesting equipment make it possible for the farmer to select almost any row configuration he chooses. The limitation today is in knowing the yield increase possible with ultra-narrow row systems, and which row spacing and population configuration results in the highest return.

Recent studies on maize and cotton at the Texas A&M Blackland Research Center on Vertisols illustrate that ultra-narrow row systems can substantially increase grain and lint yield of these dryland crops (Table 17.4). For maize, performance of the ultra-narrow rows was significantly better when rainfall was limited in 1998. For cotton, ultra-narrow rows performed equally well each year, resulting in 30–40% higher yield.

Table 17.3. Soil organic carbon to 20-cm depth after 10 years' continuous annual tillage, no tillage and native prairie in a Houston Black soil.

	After wheat (t ha^{-1})	After maize (t ha^{-1})	Native prairie (t ha^{-1})
Tilled	47.0 ± 1.1[a]	42.7 ± 0.8	66.3 ± 5.7[b]
No tillage	45.9 ± 2.1	45.6 ± 3.6	

[a]Mean ± standard deviation, $n = 4$.
[b]Mean ± standard deviation, $n = 7$.

Table 17.4. Maize grain and cotton fibre yields for different row spacing under dryland management in central Texas.

	Year		
	1996	1997	1998
Maize (kg ha^{-1}) at spacing:			
0.5 m		9909	6378
0.75 m		9527	5221
1.0 m		9209	5241
Cotton (kg ha^{-1}) at spacing:			
0.19 m	879	878	632
0.38 m	902	1042	468
0.75 m	636	764	399
1.0 m	–	–	358

Row-zone tillage systems apply tillage only to narrow strips of soil where individual rows will be planted. The tilled areas should be no more than 25% of the field area. Row-zone tillage systems provide many of the benefits of no tillage, as most of the field remains protected by crop residues, but the seed is planted in tilled areas that warm quickly and are well aerated. If tillage can be conducted in the autumn, when the soil is normally quite dry, compaction problems associated with tillage may also be avoided. Experiments are currently under way to evaluate the row-zone tillage concepts on Texas Vertisols.

References

Adams, J.E., Arkin, G.F. and Ritchie, J.T. (1976) Influence of row spacing and straw mulch on first stage drying. *Soil Science Society of America Journal* 40, 436–442.

Allen, P.M. and Maier, N.D. (1993) Growing cities: impacts of urbanization in the Blackland Prairie. In: Sharpless, M.R. and Yelderman, J.C. (eds) *The Texas Blackland Prairie: Land, History, and Culture*. Baylor University, Waco, Texas, pp. 237–251.

Arkin, G.F., Burnett, E. and Monk, R. (1978) *Wide-bed, Narrow Row Sorghum Yields in the Blackland Prairie*. Texas Agricultural Experiment Station MP-1377. Texas A&M University Printing Center, College Station, Texas.

Asamenew, G., Jutzi, S.C., Tedla, A. and McIntire, J. (1988) Economic evaluation of improved Vertisol drainage for food crop production in the Ethiopian highlands. In: Jutzi, S.C., Haque, I., McIntire, J. and Stares, J.E.S. (eds) *Management of Vertisols in Sub-Saharan Africa.* Proceedings of a conference held at ILCA, Addis Ababa, Ethiopia, 31 August–4 September 1987. ILCA, Addis Ababa, pp. 263–283.

Bouma, J. and Loveday, J. (1988) Characterizing soil water regimes in swelling clay soils. In: Wilding, L.P. and Puentes, R. (eds) *Vertisols: Their Distribution, Properties, Classification, and Management.* Technical Monograph No. 18. Texas A&M University Printing Center, College Station, Texas, USA, pp. 83–96.

Chichester, F.W. and Richardson, C.W. (1992) Sediment and nutrient loss from clay soils as affected by tillage. *Journal of Environmental Quality* 21, 587–590.

Cook, G.D., So, H.B. and Dalal, R.C. (1992) Structural degradation of two Vertisols under continuous cultivation. *Soil and Tillage Research* 24, 47–64.

Coughlan, K.J., Smith, G.D. and Yule, D.F. (1989) Soil physical research for improved dryland crop production on Vertisols in Queensland, Australia. In: *Management of Vertisols for Improved Agricultural Production: Proceedings of an IBSRAM Inaugural Workshop,* 18–22 February 1985, ICRISAT Center, India. ICRISAT, Patancheru, AP, India, pp. 87–99.

Coulombe, C.E., Wilding, L.P. and Dixon, J.B. (1996) Overview of Vertisols: characteristics and impacts on society. *Advances in Agronomy* 57, 289–375.

Freebairn, D.M., Loch, R.J., Glanville, S. and Boughton, W.C. (1984) Use of simulated rain and rainfall–runoff data to determine 'final' infiltration rates for a heavy clay. In: McGarity, J.W., Hoult, E.H. and So, H. (eds) *The Properties and Utilization of Cracking Clay Soils: Proceedings of a Symposium,* 24–28 Aug. 1981, Armidale, New South Wales, Australia. Reviews in Rural Science No. 5. University of New England, Armidale, New South Wales, Australia, pp. 348–351.

Harman, W.L., Davis, R.M., Morrison, J.E., Jr and Potter, K.N. (1996) Economics of wide-bed farming systems for Vertisol (clay) soils: chisel tillage vs. no-tillage. *Journal of the American Society of Farm Managers and Rural Appraisers* 60, 129–134.

Morrison, J.E., Jr, Gerik, T.J., Chichester, F.W., Martin, J.R. and Chandler, J.M. (1990) A no-tillage farming system for clay soils. *Journal of Production Agriculture* 3, 219–227.

Newman, A.L. (1986) Vertisols in Texas: some comments. USDA-SCS, Temple, Texas. (Mimeo.)

Potter, K.N., Torbert, H.A. and Morrison, J.E., Jr (1995) Tillage and residue effects on infiltration and sediment losses on Vertisols. *Transactions of the American Society of Agricultural Engineers* 38, 1412–1419.

Potter, K.N., Morrison, J.E., Jr and Torbert, H.A. (1996) Tillage intensity effects on corn and grain sorghum growth and productivity on a Vertisol. *Journal of Production Agriculture* 9, 385–390.

Potter, K.N., Torbert, H.A., Jones, O.R., Matocha, J.E., Morrison, J.E., Jr and Unger, P.W. (1998) Distribution and amount of soil organic carbon in long-term management systems in Texas. *Soil and Tillage Research* 47, 309–321.

Potter, K.N., Torbert, H.A., Johnson, H.B. and Tischler, C.R. (1999) Carbon storage after long-term grass establishment on degraded soils. *Soil Science* 164, 718–725.

Reicosky, D.C., Kemper, W.D., Langdale, G.W., Douglas, C.L., Jr and Rasmussen, P.E. (1995) Soil organic matter changes resulting from tillage and biomass production. *Journal of Soil and Water Conservation* 50, 253–261.

Richardson, C.W. (1993) Disappearing land: erosion in the Blacklands. In: Sharpless, M.R. and Yelderman, J.C. (eds) *The Texas Blackland Prairie: Land, History, and Culture*. Baylor University, Waco, Texas, pp. 237–251.

Seiny-Boukar, L., Floret, C. and Pontanier, R. (1992) Degradation of savanna soils and reduction of water available for the vegetation: the case of northern Cameroon Vertisols. *Canadian Journal of Soil Science* 72, 481–488.

Texas Almanac, 1998–1999 (1998) Dallas Morning News, Dallas, Texas.

US Department of Agriculture (1993) *Economic Research Service 1925–1987 Census of Agriculture Summary*. US Government Printing Office, Washington, DC.

US Department of Agriculture (1998) *1997 Census of Agriculture*. US Government Printing Office, Washington, DC.

Part IV

Conclusions

Research Needs and Opportunities for Farming Vertisols Sustainably

18

J.K. SYERS,[1] E.T. CRASWELL[2] AND P. NYAMUDEZA[3]

[1]Department of Natural Resources and Environmental Sciences, Naresuan University, Phitsanulok, Thailand; [2]International Board for Soil Research and Management (IBSRAM), Bangkok, Thailand; [3]Department of Research and Specialist Services, Chiredzi Research Station, Chiredzi, Zimbabwe

Key words: Vertisols, soil water, soil management, nutrients, cropping systems, agricultural research

Background

The International Board for Soil Research and Management (IBSRAM) and the Department of Research and Specialist Services organized a participatory workshop on the 'Sustainable Management of Vertisols in Africa' in Harare, Zimbabwe, from 8 to 15 May 1999. The workshop identified the needs and opportunities for further research, development and demonstration work on the sustainable management of Vertisols, with a primary focus on enhanced adoption by farmers.

Perhaps not surprisingly, the areas of soil surface management for improved water-use efficiency, integrated nutrient management and cropping systems on Vertisols were identified as priority issues. Technology development, validation and demonstration were regarded as being the key activities but it was recognized that alternative improved strategies were required for successful technology development and implementation.

The uptake by farmers of improved technologies for the management of Vertisols has often been erratic and usually slow. Inadequate recognition of the socio-economic constraints facing resource-poor farmers has frequently led to limited adoption. With the recognition of the value of participatory technology development techniques and the availability of new tools for facilitating the implementation of sustainable land management practices, the prospects for adoption are much improved.

In this concluding chapter the research needs and opportunities, defined at the workshop, for farming Vertisols sustainably are outlined and supplemented, where appropriate, with information from the preceding chapters.

Soil Surface Management Technologies to Improve Soil Water Management

Reducing the impact of drought through the use of appropriate soil moisture conservation practices, including water harvesting and storage, is a major need for the sustainable management of Vertisols in semi-arid areas (Hussein and Adey, Chapter 2). The development and introduction of tied furrows for moisture conservation in Zimbabwe has produced substantial increases in crop yields (Nyamudeza *et al.*, Chapter 10) but adoption has still essentially been limited to areas around the research station where the work was carried out.

Similarly, the prevention of waterlogging by the removal of excess water in the sub-humid regions by land shaping, as with cambered beds in Ghana, has given substantial increases in crop yields (Asiedu *et al.*, Chapter 11), with valuable increases in the efficiency of fertilizer use (Syers *et al.*, Chapter 3). There is a need for further research to understand nutrient dynamics in these systems, particularly at the single nutrient level, to optimize the efficiency of fertilizer use. However, the overwhelming requirement is for development and demonstration work to enhance adoption.

For both moisture conservation and moisture removal it is necessary to use implements to shape the land, whether it be the formation of ridges, furrows or cambered beds. The basic implements have already been developed, for example, low-cost animal-drawn implements in Tanzania (Astatke and Jabbar, Chapter 13), the tractor-drawn ridger in Zimbabwe (Nyamudeza *et al.*, Chapter 10), or the tractor-drawn cambered bed former in Ghana (Asiedu *et al.*, Chapter 11). Whereas further improvements will always be desirable, the availability of these implements, including tractors, to resource-poor farmers is less of a research issue than a question of policy that remains as major constraint to enhanced adoption. Jabbar *et al.* (Chapter 12) report encouraging rates of adoption by farmers of broadbed makers in Ethiopia underpinned by an improved policy environment and non-governmental organization (NGO) action programmes.

Integrated Nutrient Management

It is commonly recognized that Vertisols are inherently fertile soils and that their limitations for crop production are of a physical rather than a chemical or biological nature. Nevertheless, N remains as a common deficiency in tropical Vertisols (Syers *et al.*, Chapter 3) and there is a need for further work to refine fertilizer N recommendations, particularly in relation to moisture supply. Risk management with regard to fertilizer N use in semi-arid areas is a priority research and development need. Also, the dynamics of P in soils subjected to drying in the dry season and waterlogging in the wet season are poorly understood, especially where moisture management practices are being used, and further work is required.

The continuous removal of nutrients through crop off-take, without replenishment, inevitably leads to depletion and it is important that appropriate monitoring systems for assessing nutrient status and trends over time be in place. Nutrient monitoring is seen to be an important need by the national agricultural research and extension systems (NARES) from several countries. Modelling offers very good opportunities for assessing and predicting nutrient depletion and hence nutrient requirements.

Sustainable nutrient management requires that both organic and inorganic forms of nutrients be used as efficiently as possible. Often there are restrictions on the use of crop residues because of limited availability resulting from competing uses, particularly for animal feed. Burning of crop residues should be avoided at all costs. Similarly, animal residues returned to the land as manure are not always collected and stored efficiently, resulting in losses of valuable nutrients. Effective organic material and nutrient management systems are urgently required to minimize losses and maximize the return of nutrients in crop and animal residues. Chemical fertilizer input, especially that of N, will be required inevitably, if not now then in the near future, and this must be judicious because of the high cost to resource-poor farmers.

Salinity is not usually a problem in farming Vertisols but, where it is, salt balances need to be studied if crop yields are not to be affected adversely. This is especially important where irrigation is used. Here, effective management practices for salinity must be developed, validated and demonstrated.

Cropping Systems to Facilitate Improved Water and Nutrient Management

Crops that can be grown successfully on Vertisols are determined largely by climate, although this can be changed where irrigation is available, and by the inherent soil physical properties themselves. For example cotton is successfully grown on Vertisols (hence the term 'black cotton' soils) in many countries of the world covering a wide range of environments. Further work on cropping systems is primarily seen in the context of enhancing and retaining moisture and improving nutrient status and availability. The control of erosion by the use of an appropriate cover crop is also an important need in some situations.

Optimizing the input of organic materials by using better residue management practices has been highlighted in connection with nutrients but this is also important for increasing and/or maintaining soil organic matter levels, particularly in Africa. The use of improved fallow systems and legumes to increase N input through biological N fixation is seen as important. Also, the further screening of indigenous grasses, legumes and trees warrants further attention.

Weeds remain a key issue affecting crop production and increasing labour and/or energy requirements in several countries (e.g. Ghana, Sudan, Tanzania and Zambia). Although the use of cover crops and cultivation provides some respite, chemical herbicides provide the most effective means of control. There is a major need for low-cost, effective herbicides but in the absence of these it is necessary to integrate, as far as possible, the options for weed control to develop effective control methods.

Strategies for Future Technology Development and Implementation

The development and implementation of appropriate technologies for farming Vertisols sustainably remains a key issue, with implementation being of particular importance. Recognition of the value of participatory approaches to technology development and identification of the actors involved (King *et al.*, Chapter 15), in terms of their roles, perceived issues and expectations, is seen to be an important starting point. Improving the flow of information (see next section) and promoting a systems approach to the way in which actors contribute to technology development are two potentially valuable ways forward.

The recent development and exploitation of new tools (Craswell and Penning de Vries, Chapter 4) for implementing sustainable land management practices offer greatly improved prospects of success. These tools include the concept of resource management domains as a basis for technology sharing, which was the topic of an IBSRAM-led workshop held in Kuala Lumpur, Malaysia, in August 1996, and procedures for sustainability assessment involving the Framework for Evaluation of Sustainable Land Management, developed by IBSRAM and its collaborators. When these new tools are used in conjunction with participatory methods for identifying knowledge gaps and developing, validating and demonstrating new technologies, the prospects for more effectively and efficiently meeting the needs of resource-poor smallholders who are farming on Vertisols should be very much improved (see Fig. 18.1 for a scheme of the improved research strategy as devised by a working group at the workshop). This needs to be tested. What is required is a new research initiative that capitalizes on previous achievements and uses these new tools and approaches to conduct applied/adaptive research and demonstration on the key research issues identified.

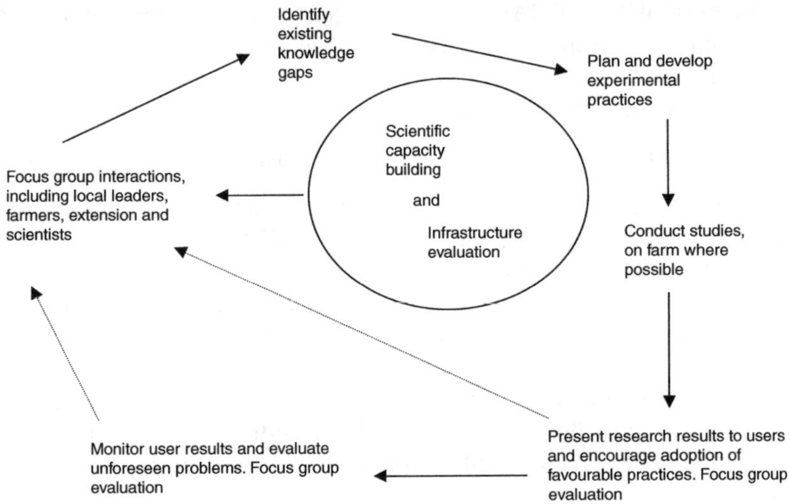

Fig. 18.1. Proposed strategy for conducting research.

Information and Networking

It has already been emphasized that there are various constraints to farming Vertisols sustainably and these relate primarily to water management, soil fertility and socio-economic problems. There have been several national and international research and development efforts to address these constraints but they have met with limited success. These efforts have often been hampered by poor coordination and lack of information exchange and dialogue by stakeholders at all levels, as well as by short-termism on the part of donors.

With a view to coordinating efforts and promoting more effective research and development on sustainable management for smallholder farmers on Vertisols, particularly in Africa, it was recommended at the Harare workshop that a consortium (VERTICON) be established.

The suggested mission statement of the VERTICON would be 'To achieve food security and poverty alleviation for smallholder farmers on Vertisols in Africa while maintaining or improving the quality of the resource base.' The VERTICON would add value to existing and future participatory sustainable land management activities by:

1. Stimulating further participatory applied/adaptive research and demonstration on priority issues for farmers on Vertisols.
2. Collating, synthesizing and evaluating published material and experimental data to produce information which would be distributed to all stakeholders.
3. Encouraging project and institutional linkages at both national and regional levels.

4. Capacity building of NARES and NGO partners, including research and extension officers, through training in the application of relevant research and extension tools.
5. Project monitoring and evaluation.

Major outcomes from the activities of this Consortium would include:

1. Expanded use by resource-poor farmers of improved soil management practices developed through participatory applied/adaptive research and demonstration.
2. Improved understanding of the socio-economic and biophysical processes essential for the sustainable management of Vertisols.
3. Enhanced capacity of NGOs and NARES to undertake the appropriate research and development work required for sustainable Vertisol management.
4. Increased effectiveness of research, development and extension efforts and increased efficiency of donor funds based on the Consortium Model.

The outputs from the project would include:

1. Protocols for the conduct of participatory technology development and demonstration work on Vertisol management.
2. A training manual for extension officers who work with smallholder farmers on Vertisols.
3. Leaflets for farmers on key issues relevant to the management of Vertisols – for example, on surface soil management, integrated nutrient management and weed control.
4. A synthesis of applied research findings on soil, water and nutrient management and cropping systems at the national and sub-regional level.
5. A natural resource database (georeferenced) that can be used at local, national and sub-regional scales.

The proposed VERTICON should preferably be organized and operated within the framework of the regional NARES and associated with related initiatives at regional and international levels. IBSRAM was seen as an appropriate lead organization for developing the VERTICON, in conjunction with NARES, sister institutions and donors.

Conclusions

Vertisols are often under-used soils that have been rather neglected in terms of research and development activities. Even though the constraints to enhanced production on Vertisols are reasonably well known, previous attempts at technology development and implementation have achieved only limited success, at least at the smallholder farmer level, and adoption of improved technologies has been less than expected by research scientists and extensionists.

With new tools for evaluating and implementing sustainable land management and with participatory approaches to technology development and implementation, the prospects for conducting more effective research and development are much improved. A consortium (VERTICON) is proposed to add value to existing and future technology development work on priority issues relating to the sustainable management of Vertisols. Through the activities of the VERTICON, the prospects for farming Vertisols sustainably will be substantially enhanced.

Index

Page numbers in *italics* refer to figures and tables

Browse Read and Buy
www.cabi.org/bookshop
ANIMAL & VETERINARY SCIENCES
BIODIVERSITY CROP PROTECTION
HUMAN HEALTH NATURAL RESOURCES
ENVIRONMENT PLANT SCIENCES
SOCIAL SCIENCES
Subjects
Search
Reading Room
Bargains
New Titles
Forthcoming
CABI Publishing
A division of CAB International
Online BOOK SHOP
Crop Pollination by Bees
Keith S. Delaplane and Daniel F. Mayer
A DICTIONARY OF Entomology
G Gordh and D H Headrick
Order & Pay Online!
MasterCard
VISA
AMERICAN EXPRESS
Principles of CATTLE PRODUCTION
C.J.C. Phillips
Seeds
THE ECOLOGY OF REGENERATION IN PLANT COMMUNITIES
2nd EDITION
Edited by Michael Fenner
FULL DESCRIPTION
BUY THIS BOOK
BOOK OF THE MONTH
Tel: +44 (0)1491 832111 Fax: +44 (0)1491 829292